Michael

THE REALLY PRAC GUIDE TO PRIMARY GEOGRAPHY

Second Edition

Marcia Foley and Jan Janikoun

Stanley Thornes (Publishers) Ltd

Acknowledgements

The authors would like to thank the following people for their help, support and forbearance during the writing of this book: David Carter, Mike Foley, Graham Matthews and Lynne Dixon.

The publishers would like to thank: Sandling CP School, Kent; Alan Waters, Cumbria; Steve Telfer and the Geographical Association for permission to reproduce charts on pages 34, 54, 59, 61, 65 and 99. Efforts have been made to contact copyright holders and we apologise if any have been overlooked.

First published in 1992 by:
Stanley Thornes (Publishers) Ltd
Ellenborough House
Wellington Street,
CHELTENHAM GL50 1YW
England

Second edition 1996
98 99 00 01 / 10 9 8 7 6 5 4 3

A catalogue of this book is available from the British Library.

ISBN 0 7487 2617 9

Typeset by Northern Phototypesetting Co. Ltd, Bolton
Printed and bound in Great Britain at Scotprint

Contents

How to use this book

Our aims when writing this book were three-fold:

- To help teachers become aware of all aspects of primary geography
- To promote good geographical practice in primary schools
- To help teachers develop National Curriculum geography.

It has been written for a wide professional audience: teachers, headteachers, students in initial teacher training, those returning to the profession and other interested LEA colleagues.

We have tried to make the text as practical as possible, dealing with the various aspects of planning and implementing primary geography in as much depth as space constraints would allow. We do not envisage that you will read the whole book from cover to cover in one sitting. It can be read as a whole but has also been designed so that individual chapters can be read separately if you wish to inform or update yourself on a particular aspect of geography and curriculum planning. Cross references are given so that you know to which other chapters you might need to refer.

Geography is the focus of this book, but we recognise that it is but one part of the whole primary curriculum. Nevertheless we have presented a full model for planning, developing and implementing geographical learning in the primary school. By highlighting the nature of geography we hope you will come to a better understanding of the subject that will help you develop geography in your school, whatever your role.

Schools are at many different stages of development with primary geography. Each school's needs are unique but we hope that within the range of the book you will find clarification and practical assistance – whether you are beginning to think about primary geography or refining your expertise. We have used the term 'coordinator' throughout the book. We recognise that this might be one individual with just that curriculum responsibility; it could be somebody looking after a number of subjects; or a curriculum team. We also recognise that smaller schools might find it difficult to identify with any of these models. The important idea is that there is a way of organising planning and support for colleagues as geography develops.

The current notation for the Revised English National Curriculum geography programmes of study is used throughout this book. For example, KS2 PoS 5b indicates that in key stage 2 pupils should be taught about 'how localities may be similar and how they may differ.'

The Welsh Order programme of study numbers are slightly different and there is more variation in the themes in key stage 1, and some locality and theme differences in key stage 2. Readers needing to use the Welsh Order are advised to refer to it for details. The same good practice, planning and assessment principles apply to the Welsh Order as the English one which is referenced here.

Everything in this book has either been done by ourselves whilst teaching, or developed for, or during, in-service training, or from working alongside colleagues in and outside classrooms. So we would like to thank all those teachers and children with whom we have had contact and who have shared our thinking and helped us develop our ideas.

1

GOOD PRACTICE IN PRIMARY GEOGRAPHY

What is geography?

Geography is the study of people and places, and the interaction between them. All primary teachers teach some geography even though they may be surprised to realise it; many have had little formal geographical education beyond the age of 13 or so and therefore do not consciously recognise the subject matter. Those of you who teach geographical matters well (whether as part of a broad topic or as a specific geography topic), do so because you are able to recognise geographical concepts and skills and thus the subject's contribution to children's learning.

Geography is not merely the ability to memorise the names of places and locate their position on the earth's surface. Knowing the name and location of places and features (sometimes referred to as 'capes and bays geography') is an integral but small part of a much larger and more fascinating area of learning. Geography consists of knowledge and understanding of concepts and skills, all of which relate to the physical and human environment and the interaction between them. Geography is also an enquiry-based subject and any questions raised whilst teaching geography will be related to one of a number of key areas of enquiry.

Key geographical questions

There are seven key questions which are fundamental to children's learning about geography:

- Where is this place?
- What is this place like?
- Why is this place as it is?
- How is this place connected to other places?
- How is this place changing?
- What is it like to be in this place?
- How is this place similar to, or different from, another place?

In order to teach and learn about geography, it is good practice to bear in mind these key questions. Asking children about any place brings out the essential learning about it, be it a feature – such as a mountain, a quarry, a village, a country – or a natural region, such as a desert. If the use of key questions is new to you, then Figure 1.1 gives some examples of how the answers to such questions can develop key geographical concepts and vocabulary.

Working through a key question approach helps children build up an understanding of processes, places, and the patterns they make. It provides a framework of knowledge in which to develop key geographical concepts.

Figure 1.1

Key geographical concepts and questions

Question 1: Where is this place?
Question 2: What is this place like?
helps develop the concept of location and a sense of place.

'My village is in the country, on a hill top, in East Sussex, England. It is called Rotherfield. It is small with a food store, a post office shop, a school, a garage and some pubs ...'

Question 3: Why is this place as it is?
helps develop the concepts of spatial pattern and process.

'... the village is difficult to drive through and dangerous to walk through because the main streets are narrow. They cannot be widened because the old houses, some of them very old indeed, do not have front gardens ...'

Question 4: How is this place connected to other places?
helps identify relative location and build up the idea of spatial patterns.

'... the railway was closed a long time ago, so to travel to London my dad must drive to the nearest town. There are a few buses every day, so it is not convenient for my mum if she wants to go to the supermarket ...

Question 5: How is this place changing?
helps identify changes occurring in patterns, process and systems; can bring up issues to be discussed.

'... they built a new estate of thirty houses and now there is more traffic turning onto the main road from it. But there are more children for our school, too ...'

Question 6: What is it like to be in this place/live here?
develops the concept of place and helps us to think about attitudes and values.

'It's good! I like it. We can play on the recreation ground opposite my house.'

Question 7: How is this place similar to/different from another place?
develops the concept of similarity and difference, as well as location and place by comparison.

'... it's smaller and quieter than where I lived before in the suburb of a big town. But my friends don't all live here so I have to rely on them coming to see me, unless it's the weekend when I can travel to see them by car.'

Geographical concepts

The key concepts of geography are:

- Location and place
- Spatial pattern
- Process
- Systems
- Similarity and difference.

Asking children key questions when they approach geographical work helps them progress towards the ultimate understanding of key geographical concepts.

Figure 1.2 shows how particular key questions relate to the first four concepts, and how each concept can be broken down into sub-concepts.

The final concept, similarity and difference, relates to all the other aspects of geography. In learning about different places, patterns, processes and systems children should be encouraged to compare and contrast so that they build up a more complex knowledge and understanding of place and space.

The important question which children can ask – What is it like to be in this place? – relates to attitudes, values and issues.

Figure 1.2

Key questions help develop key concepts		
Question	**Main concept**	**Subconcepts**
1 Where is this place?	Location	Relative location: where is it, related to another place?
2 What is this place like?	Spatial pattern Location	Patterns made by features in the physical landscape Patterns made by features in the human landscape Distribution of features
3 Why is this place as it is?	Spatial pattern Process Systems and location	Inputs, outputs, interaction, routes
4 How is this place connected to other places?	Locations Systems Pattern	Position, relative location Networks
5 How is this place changing?	Process Systems	Change over time: development, rural, urban Cycles (rock, weather, water) Interaction
6 What is it like to be in this place?	Place/environment	Leads to understanding attitudes, values, beliefs within issues
7 How is this place the same as, or different from, another? Extensions/Overarching question	Similarity and difference	Assumes comparisons based on above six questions, relating to two or more places

Attitudes, values and issues

The people who live in any place have feelings about its features, and attitudes towards the processes and systems that arise from their set of values. They also have attitudes towards local issues. By asking primary children 'What would it feel like to be in this place?' after they have some knowledge and understanding of it, we can help develop their understanding of the range of attitudes, values and beliefs contained in any issue. We are asking pupils to try to put themselves as far as possible in the situation of the people and places they are studying.

As soon as children begin to consider this question, they may ask: 'How are decisions made about this place?', questioning the concepts of people's rationale and motives. 'Why has the factory been built here, when some residents were against it?'

Once questions about decision-making arise, an opportunity to build on an issue for enquiry may result. With primary children the issue will usually be a local one, as first-hand investigation or fieldwork will make the issue vital. However, it is possible to generate excellent work on issues through role-play activities. Hypothetical issues avoid the sensitivity of local issues in which pupils' parents may be involved.

Some teachers shy away from dealing with issues in geographical work because they immediately raise questions about children's and adults' attitudes and values. Their own attitudes and values may be questioned, too. Good teachers recognise however, that good geographical work cannot avoid dealing with such matters. Children need to know and be able to accept that society is so complex that people do not always come up with 'right' or 'wrong' answers. Most solutions are a compromise and will go against some people's wishes whilst fulfilling those of others. The DES HMI series Matters for Discussion, number 5, *The Teaching of Ideas in Geography 1978* states that:

> "*Geography offers the opportunity of situations where responsible efforts can be made to help pupils understand the nature of values and attitudes and their importance in making decisions.*"

This is not a matter of indoctrination but of raising the possibility of helping children from a young age independently to make some sense of the world around them. Although the influence of the school is slight compared with that of the home environment, media and society in general, primary teachers through geography, can help children to develop:

- A respect for evidence
- An awareness of biased information and intolerance
- An awareness that simple explanations rarely tell the whole tale
- An interest in other people and places
- Empathy with other life styles and cultures, including minority groups in the United Kingdom
- A concern for the quality of rural and urban environments
- The ability to appreciate other points of view and to reach compromise solutions
- The concern to value and conserve resources.

Dealing with attitudes, values and issues in geography contributes to pupils' moral education (see Chapter 11).

The enquiry process

Good geographical teaching and learning uses the enquiry process to motivate children to find out about the physical and human environment and their interrelation. We can pare down these seven key questions depending on what we are finding out about.

What does the enquiry process involve?

HMI explain that:

> "*Pupils should not be primarily passive recipients of information, but should be given adequate opportunities to carry out practical investigations, to explore and express ideas in their own language ... and to reflect on other people's attitudes and values.*"
>
> DES, *Geography from 5 to 16* (HMSO, 1986)

National Curriculum guidance issued in connection with the original Geography in the National Curriculum Order explains enquiry as follows:

"Enquiry is the process of finding out answers to questions. At its simplest, it involves encouraging children to ask questions and search for answers, based on what they might already know and from data sources. As their skills develop, children can move to a more rigorous form of enquiry involving the development and testing of hypotheses."

NCC, *An Introduction to Teaching Geography at Key Stages 1 and 2* (NCC, 1993)

The 1995 Revised Geography Order places geographical enquiry firmly in the PoS for both key stage 1 and 2 by listing key questions in PoS 1 and highlighting the enquiry process in PoS 2. For example, in KS2 PoS 2:

"In investigating places and themes, pupils should be given opportunities to:

a observe and ask questions about geographical features
b collect and record evidence to answer these questions
c analyse the evidence, draw conclusions and communicate feelings."

DFE, *Geography in the National Curriculum* (HMSO, 1995)

How do we go about the enquiry process?

Figure 1.3 shows examples which could be used with infants and juniors, depending on the sophistication of data collected. At key stage 1 the context is the school grounds, whilst at key stage 2 it is the local area.

How many questions do we ask or expect children to formulate for an enquiry?

There is a temptation to think that many questions must be asked. However, it is sufficient to ask one question only – the age and experience of the children; the experience of the teacher in using the enquiry process and the time available can be the only guidance for the teacher's professional judgement.

Figure 1.4 lists further questions and statements which can trigger good geographical work. These questions are suitable for key stages 1 and 2, depending on the level to which they are developed.

The content of geography

Traditionally geography has two themes: the physical environment and the human environment. Figure 1.5 suggests the way in which these two central elements interact with each other and Figure 1.6 expands these elements.

The Teaching of Ideas in Geography, from the DES HMI 1978 series, Matters for Discussion, details the breakdown of the areas listed above for the specialist. Figure 1.6 indicates in straightforward terms the kind of content with which each theme might deal at primary geography level.

This is not the only way of describing the content of geography: there are many others. The geography Statutory Order represents a very specific way of describing content with an emphasis on places, themes and skills.

Environmental geography appeared as a chunk or 'attainment target' in the 1991 Geography Statutory Order. This was consistent with the raising of consciousness about the environment in general during the 1980s.

The 1995 Geography Statutory Order includes environmental quality and change for key stages 1 and 2. Thus environmental work is still a compulsory part of the geography curriculum.

Nevertheless, specialist geographers see environmental awareness as a life skill, and

Figure 1.3

Enquiry process in theory	Enquiry process in action in school grounds KS1	Enquiry process in action in a local area KS2
Recognise an issue or focus for enquiry	Where shall we have a compass rose sited in our school grounds?	A large new pub is to be built on a local derelict site.
Ask some relevant questions or make a statement to be investigated (one or several as appropriate)	Should it be painted on tarmac, or bricks on the grass? In an open or exposed site? Near our classroom or not?	Is this the best use for the site? What services and leisure activities already exist in our place? Are people happy with these?
Collect relevant data	Check proposed sites for sun/shade – take wind, temperature judgements (for infants), measurements (for juniors), etc. Check sites for flatness, drainage. Do a 'route' survey – would the proposed sites be in the official rights of way - on the football pitch, etc?– on the shortest route from hall to another building, etc?	Make up questionnaires to address these questions. Questionnaires to parents/other classes/local residents depending on time constraints. Visit locality to survey and plot leisure facilities and services on map. Write to/visit/invite into school (according to time constraints, etc.) Planning Department official, brewing company representative, etc.
Interpret and analyse data	Using data collected begin to decide on best location for compass rose.	Using data collected (via IT data base if possible) work out whether survey is in favour of new pub.
Present findings	Use plans, maps, diagrams to compare pros and cons of various sites.	Examine alternative suggestions and reasons. Produce 'public enquiry' role play to illustrate findings (this could be presented to another class, in a school assembly or before the class's parents).
Draw conclusions	Decide on final siting – propose site to headteacher	Write to local planning department with results of findings.
Evaluate enquiry	See how well the actual site works out in practice – is it accessible, not too exposed, etc? This would be monitored over a term or a year!	Evaluate what was learned about local area.

essential to an individual's development and sense of place. Geography is about physical and human environments and the interaction between them. Thus the environmental aspect of geography is an inherent part of the subject, not a 'bolt-on' extra in the form of a cross-curricular dimension to be delivered through the subject.

Geographical skills

Mapwork is a major geographical skill. Indeed, to many people both inside and outside the world of education, mapwork *is* geography! However, this view is a very narrow and inaccurate one. The making,

Figure 1.4

Enquiry questions

Where shall we have a compass rose painted on our playground?

Where is the best place to site the new flower bed/seats/trees in our school grounds/park?

Where are the traffic or crossing danger points in our local streets/my route to school and how could we suggest improvements?

How can we improve our school grounds?

What kinds of local/town shops do we have? Are they the same as those in place X we will visit soon?

Are our nearest houses made from mainly natural or man-made materials?

How will the by-pass/roundabout/road improvements change our village/town?

How is a new road built and why is it being built?

Why is the new supermarket sited here?

What is the best shop/new building(s) to put on our nearby derelict site?

How many routes are there from place A to B (within school grounds as well as in a town)?

Why are the goods in the supermarket laid out in the order they are? Is it planned or random?

Is our town/village a tourist attraction? Why is the locality we are visiting a tourist locality?

Can we make up a barrier trail for our class friends to follow?

Can we plan a tourist trail for our village?

Can we make up a building materials trail for our village/area of town?

How have humans changed/influenced the course the stream takes?

How are many visitors to this beauty spot/site of special scientific interest, etc. affecting it?

Figure 1.5

The interaction between the physical environment and the human environment

Physical Environment		**Human Environment**
Land forms	⟷	Farming
Natural resources		Industries
Soils and vegetation		Population
Weather and climate		Recreation
		Settlement
		Towns and cities
		Transport

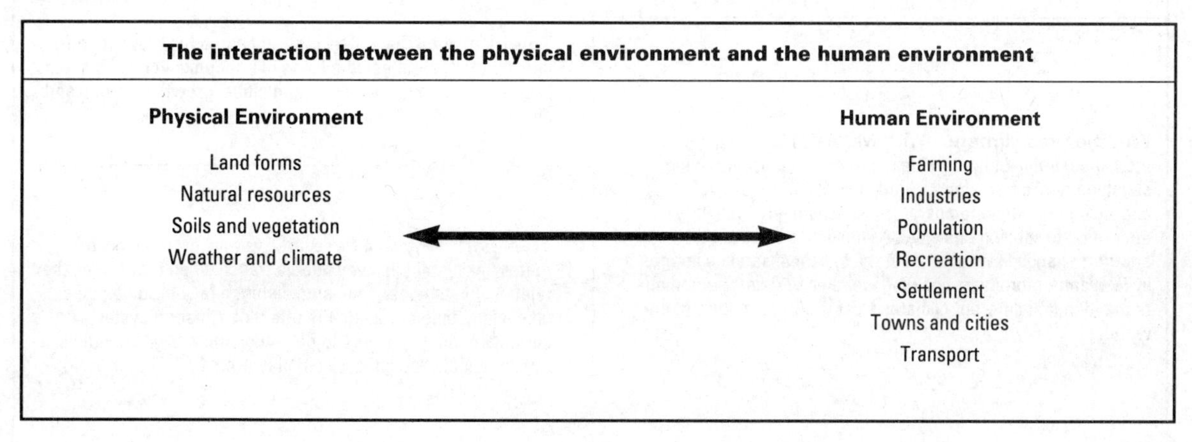

Figure 1.6

Elements of the physical environment	Elements of the human environment
Landforms Scenery, landscape (what makes the landscape) – hills, plains, rivers, mountains – and how they were formed; the rocks which form the landscape (igneous, sedimentary, metamorphic); the water cycle; outline explanation of weathering of rocks and building stone by rain, acid rain and the effect of wind; the erosion of rocks by sea, rivers, glaciers; the making of new land by the deposition of sediment / pebbles from the sea or rivers.	**Farming** Farming as an industry to provide us with food; types of farming (sheep, dairy, arable, etc.); the farm as a system with inputs (seed, animals, etc.) and outputs for sale (crops, milk, wool, animals, etc.); special types of farming and their location, e.g. pick-your-own fruit farms in heavily populated areas; how products travel to market; contrasting farming methods between UK and the developing world.
Natural resources Understanding what natural resources are: water, rocks to be quarried, minerals, etc. to be mined, forests, woodland, indigenous animals, etc.; alteration of the landscape by extraction of natural resources, e.g. reservoirs, land fill, gravel pits, lakes; how we can preserve our resources; damage and pollution of our environment by using resources, e.g. rubbish tips, waste disposal, recycling.	**Industry** Primary industries (mining, fishing, quarrying), secondary industries (steel works, food processing), tertiary industries (service industries: transport companies, travel agencies); siting of industries.
	Population People cluster together or space out to live according to the physical geography, lifestyle and employment opportunities; people migrate because of jobs, war, famine, etc.; too many people living in one place can strain resources.
Soils and vegetation Different types of soil (clay soil, sandy soil, etc.); variation in soil particle size (i.e. structure); how water drains through soil; different types of natural vegetation (rain forest, deciduous, coniferous); beginnings of knowledge of natural vegetation zones (desert, polar, tropical rainforest).	**Recreation** Leisure activities and facilities; holidays; environmental issues: increased recreation can cause pollution damage to landscape and wildlife.
	Settlement Reasons for siting of settlements (often rooted in local history): ease of communication, types of economic activities found in a place (function), the way the place has grown (form), how the place is linked to nearby places.
Weather and climate What weather is (sun, rain snow, etc.); measuring and recording weather (temperature, wind, sunshine rain, air pressure, cloud cover); the water cycle; microclimates – how temperature, sunshine, etc. can vary according to site, e.g one side of a building compared to another; change in weather over the seasons; satellite images in TV weather forecasts showing weather systems; beginnings of the idea that different climates exist in different parts of the world.	**Towns and cities** The difference between town centre and suburb (i.e. zones), the kinds of economic activities found in centres; travelling in towns and cities; growth of towns and cities.
	Transport Types of transport, systems or networks of routes linking places, why people travel, when they travel, the relationship between time and distance (a short distance can take a long time to travel), the effect of transport systems on our environment, barriers to communication, e.g. mountains, river channels, 'no go' areas in playground.

using and reading of maps is an essential and very important geographical skill, and one which is so wide that it can be broken down into many sub-skills (see Chapter 8). But mapwork is just one area of many geographical skills. These skills can be classified as primary and secondary skills, or direct and indirect skills.

Primary or direct skills involve children's first-hand experience or active learning either inside the classroom, or outside it (fieldwork) where the pupil is directly engaged in collecting data, be it information, evidence or samples. Indirect or secondary skills involve the pupil in collecting data second-hand, for example from information books, the teacher, educational television, radio or computer databases. Data and information collected via these skills then have to be handled, analysed and presented in some form which involves further skills, such as:

- Producing a map
- Producing a piece of writing
- Making a pie chart
- Constructing a line graph
- Making a flow chart
- Performing a role play.

The representation of geographical information in a non-written form, for example in charts, maps, diagrams or graphs, is referred to as graphicacy. The various groups of skills essential to good practice in primary geography will be examined in more detail in Chapters 8, 9 and 11 along with progression in the acquiring of these skills.

Geographical vocabulary

Each area of learning has a vocabulary which, although part of the English language as a whole, is also specific to its own area of learning. Geography has its own particular vocabulary which children begin learning unconsciously as soon as they speak. They become familiar with geographical words, build up their range of vocabulary and are able to use them in geographical contexts, for example 'this way', 'that way', 'map', 'signpost', 'shop', 'doctor', 'town', 'river', 'temperature', 'cloud', 'by-pass', 'country', 'equator', 'soil', 'desert', 'quarry', 'mountain'.

If you are aware that much of the vocabulary that you and your pupils use has a geographical context, then you will be able to encourage children's knowledge and use of it. Continuity and progression in geographical vocabulary is part of good practice in primary geography. It is possible to identify chains of vocabulary which relate to the content of geography, for example home – house/farm – village – estate – suburb – town – city – capital city – settlement. A seven-year-old would happily use 'home', 'house', 'farm', and maybe 'town' or 'village', but it would only be reasonable to expect an older junior to use 'settlement' in context instead of 'lots of houses grouped together to make a town or village'.

The word chains suggested in Figure 1.7 give an idea of the range of words which primary children should be familiar with by the time they leave. It is possible to ask juniors to come up with their own word chains based on geographical vocabulary – it generates much discussion in groups! Such lists are illustrative, not exhaustive. Vocabulary in National Curriculum geography programmes of study will be referred to in Chapter 2 and in the Glossary at the end of this book.

Figure 1.7

Physical environment word chains	
Scenery	bank – mound – mount – hill – foothills – mountains
Slopes	flat/level – plain – gentle – steep – cliff
Rock	clay – sand – silt – grit – pebbles – stones – boulders
Wind	calm – breeze – gale – storm – hurricane
Precipitation	mist – fog – drizzle – rain – downpour – hail – snow
Water	puddle – pool – pond – lake/reservoir – sea – ocean
Vegetation	tree – copse/coppice – wood – forest
Human environment word chains	
Place of worship	shrine – chapel – church – cathedral/synagogue/mosque/temple
Shopping/Retailing	market stall – shop – market – department store – supermarket – shopping centre – hypermarket
Recreation	garden – playground – playing field – recreation ground – park – theme park – countryside park – National Park
Population	few – sparse – spread out – scattered – clustered – concentrated
Environmental quality	polluted – spoilt – damaged – uncared for – well looked after – well conserved

Summary

What, then, is good practice in teaching and learning geography in the primary school all about?

It comprises:

- Teaching and learning through key questions to build up conceptual development
- An awareness of the general content of physical and human geography and how the two interrelate
- Developing geographical skills
- Using the enquiry approach
- Integrating fieldwork throughout primary children's work as a skill and through the enquiry process
- Developing children's knowledge and use of geographical vocabulary
- Dealing openly and sensitively with issues, attitudes and values.

Progression, coherence and continuity in all these elements of primary school geography are essential, whether the subject area is taught as an aspect of a broad topic or project, or as a specific subject in its own right. It also assumes that the teacher's management and organisation skills are good enough to enable teaching and learning to take place in a variety of styles – class, group, individual, teacher-directed, child-organised, and so on.

Geography is a way of studying the world and its people as they are today. With good teaching practices, the obvious relevance and usefulness of the subject to the children should make it one of the more fascinating and motivating curricular areas.

National Curriculum geography is helping many primary teachers towards the first essential step in good teaching and learning of primary geography – recognition of what the subject is about and how it can illuminate primary education. The 1995 Revised Statutory Order has the potential to develop and support good practice in teaching and learning geography in the primary school. Its potential will only be realised if teachers know how to interpret it and plan with it.

2 GEOGRAPHY IN THE NATIONAL CURRICULUM

The background

In their report *Aspects of Primary Education: The Teaching and Learning of History and Geography* (1989), HMI stated that:

> "*Although there was some good work, overall standards in geography were not satisfactory.*"

When geography was good it included:

- A curriculum coordinator in post
- Being part of topic work that achieved a balance across subjects
- A collegial approach to curriculum development
- Clear aims and objectives
- Good quality discussion and questioning in the classroom
- Use of the local environment.

However, HMI also highlighted many problems in primary geography teaching:

- Lack of physical geography
- Children often repeating topics
- Not enough use of atlases and globes
- Limited work on other places in the UK
- Little work beyond the local area
- Geography used to practise skills in English and art
- Few schools with a teacher responsible for geography
- No clear rationale for the choice of topics
- Little effective planning
- Not enough time allocated to the subject
- Curriculum often television-led
- Barely adequate resources
- Too many case studies
- Lack of continuity and progression.

This list paints a very dismal picture of primary geography. The introduction of the National Curriculum in 1991 provided an opportunity to put many of these things right. It offered schools a framework to build a curriculum that gave geography its full status as a foundation subject within the total primary curriculum. The framework was designed to be flexible so that schools could develop their own schemes of work, integrating geography, if applicable, with other subjects.

Three years after the introduction of National Curriculum Geography, the Office for Standards in Education (OFSTED) stated amongst its main findings that:

> "*In the primary schools the standards of achievement were satisfactory or better in 86% of lessons.*"

OFSTED, *Geography: A review of inspection findings 1993/94* (HMSO, 1995)

In the same report, OFSTED further identified these positive points:

- Some of the best teaching, where it was good, was to be seen in key stage 2.

- Local area work was thorough.
- Increasing resources for geography had benefitted standards of achievement and the quality of learning.
- Structured programmes of local and distant fieldwork visits, some residential, promoted good standards of work.

However, the report also identified these weaknesses:

- Standards of achievement, quality of learning and quality of teaching were weaker overall in key stage 2.
- Schemes of work to track continuity, consistency and progression were lacking.
- Policy documents were too generalised.
- Assessment was undeveloped.
- Recording and reporting needed attention.
- Places, themes and skills were insufficiently integrated in planning and teaching.
- Distant place work was too superficial.
- There was still a lack of resources for geography in some schools.
- Management and organisation of the subject left much to be desired.
- Great variability in the quality of practice, both within and between schools, was identifiable.

Comparison of these two reports which evaluate primary geography before and after the introduction of National Curriculum Geography shows many similarities in strengths and especially in weaknesses.

At first sight we may question just how much progress has been made in the intervening years! However, the key must lie in the comparison of standards in geography between the two times, where an improvement, from unsatisfactory overall in 1989 to more generally satisfactory or better in 1993/94, must highlight the progress made. Moreover, all pupils now have a legal entitlement to a statutory geography curriculum and we must continue to work to improve the standards and quality of that curriculum.

It must also be remembered when comparing the lists of weaknesses that the goal posts were moved by the National Curriculum and judgements are now made against much tighter published criteria than in 1989, both in the form of the Geography National Curriculum Order and the OFSTED framework for the inspection of schools.

The first National Curriculum Geography Statutory Order (1991) with its five attainment targets and programmes of study for these, assessed against a plethora of very specific statements of attainment, raised the status of the subject in primary learning but created the following problems:

- A lack of understanding of how to plan or interpret the requirements, due to their complexity and the use of subject specific language
- A separating out in teaching of physical, human, environmental geography, the various places required and map skills
- A highlighting of the lack of teachers' subject knowledge in key stage 2
- Time manageability issues – could it/couldn't it all be done?

The revision of this first Order now presents teachers with a much more realistic and comprehensible document.

What the Revised Order covers

Geography in the National Curriculum (1995) is now composed of one holistic attainment target made up of places, themes and skills for both key stage 1 and 2. A separate programme of study for each key stage specifies these, as summarised in Figure 2.1.

Figure 2.1

The structure of geography in the National Curriculum, 1995		
	Key stage 1	**Key stage 2**
Summary	Investigation Physical and human features Local area	Investigation Places and themes Range of scales
	Key questions Fieldwork Skills, places, themes Wider geographical context	
Skills	Geographical enquiry Vocabulary Fieldwork Making and using maps, plans, globes Using secondary sources and IT	
Places	School locality Contrasting locality (UK or overseas)	School locality Contrasting locality in the UK Contrasting locality in developing country UK context EU context Global context
Thematic studies (Themes)	Environmental quality	Rivers Weather Settlement Environmental change

NB The Revised Order for Wales differs slightly to the above. In key stage 1 *Weather* and *Jobs and journeys* are placed alongside *Environmental quality* and one of the three must be chosen.

In key stage 2, *Economic activities* replaces the British *Settlement* theme.

Figures 2.2 and 2.4 show a further breakdown of the content of these programmes of study. These may both be photocopied (see pages 205–6) and used as audit and planning check lists (see Chapters 3 and 4).

It is helpful to know and understand the structure of the 1991 Geography Order in order to comprehend the places, themes and skills principle.

Figure 2.2

Key stage 1: Programme of study elements	
1a Investigate physical and human features 1b Geographical questions: where? what? how? 1c Broader geographical context	3f Use secondary sources
2 Enquiry: observe, question, record, communicate	4 School locality: school building School locality: school grounds School locality: local area Contrasting locality (UK or overseas)
3a Use geographical terms 3b Local fieldwork 3c Follow directions 3d Make maps, plans with signs and symbols 3e Use globes Use maps and plans Identify key geographical features Locate/name countries of UK Locate home area on map Follow a route	5a Main physical and human features of localities 5b Similarities and differences between localities 5c Effects of weather on people and places 5d Land and building use 6a Likes and dislikes of quality of environment

The 1991 Order comprised five aspects or attainment targets and these were:

1 Geographical Skills
2 Places
3 Physical Geography
4 Human Geography
5 Environmental Geography.

The last three aspects were eventually referred to as geographical 'themes'. Thus the elements of Geography in the National Curriculum have been abridged to places, thematic studies and skills, or places, themes and skills. These aspects of geography interact, as Figure 2.3 demonstrates.

These different aspects of geography were not supposed to be dealt with separately but in an integrated way. For example, in key stage 2 you might study the land use (human geography) of suburban Cairo (a developing locality), while referring to the River Nile and its wider course (physical geography), using secondhand resources (a video and supporting photo pack and written information (skills)) to do so. (See Figure 2.5.)

Figure 2.3

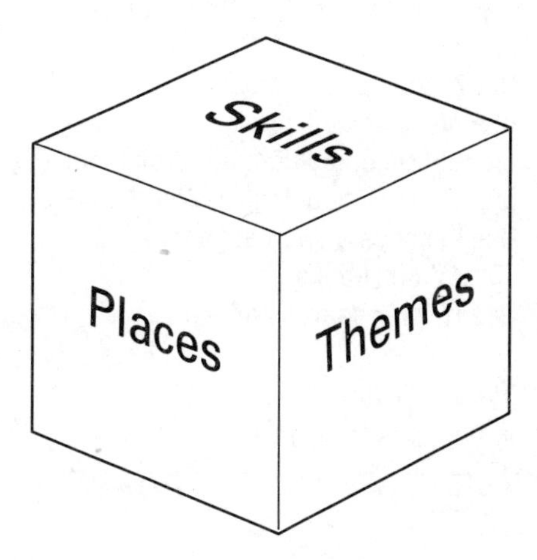

In key stage 1 you might go on a field visit (skills) to your nearby (local area) beach with its cliff and man-made sea defences (physical and human geography) and consider whether the area is clean and a good place for tourists to visit (environmental geography). (See Figure 2.6.)

Figure 2.4

Key stage 2: Programme of study elements

1a Investigate places and themes
1b Geographical questions: what? where? how and why?
1c Development of ability to recognise pattern, application of knowledge and understanding
1d Wider geographical context: links, range of scales

2a Enquiry: observe and question
2b Collect and record evidence
2c Analysis, conclusions and communication

3a Use geographical vocabulary
3b Fieldwork: instruments, techniques
3c Make maps and plans with symbols and key
Use symbols and key
3d Use globes, maps, plans
Co-ordinates
Four-figure grid references
Measure distance and direction
Follow routes
Use contents and index page of atlas
Identify specified details: UK map (A)
Identify specified details: Europe map (B)
Identify specified details: World map (C)
3e Use secondary sources pictures, photographs, TV, visitors, books
3f Use IT for research and enquiry

4 School locality
Contrasting UK locality
Contrasting locality in Africa, Asia (not Japan), South or Central America

5a Main physical and human features
5b Similarities and differences between localities
5c Effect of features of locality on human activity
5d Change in the locality
5e The broader context and links with other places

6 UK context
EU context
World context
Local scale
Regional scale
National scale

7a River systems: features and catchment area
7b River processes: erosion and deposition

8a Weather: microclimate
8b Seasonal weather patterns
8c Weather round the world

9a Settlement: characteristics, location, economic activities
9b Land use in settlements, jobs
9c Land use issues

10a People's effects on the environment
10b Managing and sustaining the environment

Unfortunately, the 1991 Order was not able to state this, either diagrammatically or in words, even though the three-dimensional nature of the subject had been stressed by the Geography Working Group in their Final Report (1990). Non-specialist teachers, and indeed many specialist teachers, proceeded to teach the subject in a very fragmented way, rather than using the combination approach described above.

The 1995 Geography Statutory Order for key stages 1 and 2 organises its programmes of study under a general paragraph 1, which acts as a kind of cross referencing summary to all other parts of the programme of study which then follow on, listed under:

- Geographical skills – programme of study paragraphs 2 and 3

Figure 2.5

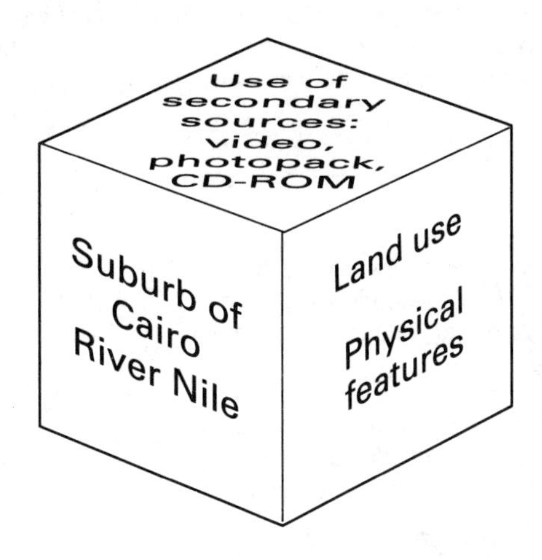

Figure 2.6

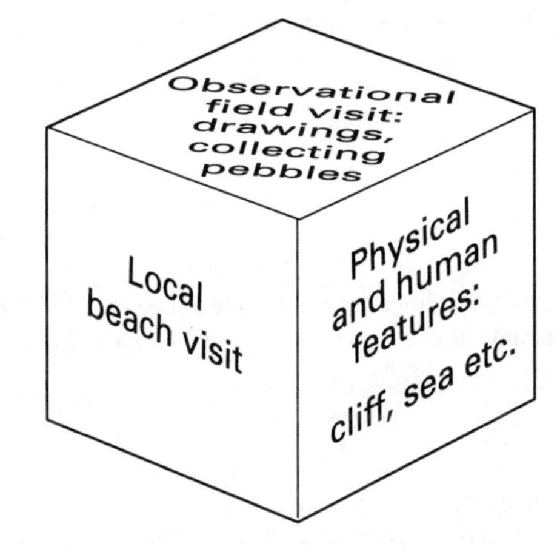

- Places – programme of study paragraphs 4 and 5 (and partly 6 in key stage 2)
- Thematic study – programme of study paragraph 6 (key stage 1) and programme of study paragraphs 6, 7, 8, 9, 10 (key stage 2).

The programmes of study are set out in numerical order on the pages of the Revised Order; there is, therefore, still a tendency for teachers planning work to do so in a fragmented way, giving importance to the *order* in which the words are written, rather than giving *equal* status to the varying aspects of places, themes and skills and planning in an integrated way.

The key stage 3 programme of study is similarly organised and it is important that geography coordinators and teachers read and/or refer to it to gain an overview of the attempt at continuity and progression that is built into the Order. It is also important for them to check that they are not spending too much time on, say, local coastal features in great depth, when they will be dealt with in depth at key stage 3.

The programmes of study are at pains to stress their integrated nature in their very wording. For example, KS1 and KS2, PoS 1b stresses:

> "*Studies should involve the development of skills, and the development of knowledge and understanding about places and themes.*"

In key stage 2, PoS 6 is a crucial and clever paragraph which is often read too quickly and whose significance is often missed. This paragraph gives directions on how themes should be investigated and the range of scales to be investigated (see Figure 2.7).

Similarly, this paragraph expands the range of places which can and need to be addressed beyond the localities already listed in Figure 2.1 and clearly highlights, yet again, that themes can be integrated with place studies across the whole range of scales (see Figures 2.8 and 2.9).

It is essential to note that geographical enquiry is a key feature of all the key stage programmes of study and should therefore permeate children's geographical learning (PoS 2).

Of the other geographical skills in PoS 3, some are very open. For example, KS1 PoS 3 'Use geographical terms' and KS2 PoS 3a,

Figure 2.7

You may study PoS 7, Rivers, and PoS 10, Environmental Change, together, for example the long-term effects of the building of the Aswan Dam on the Nile, tourist cruise boats on the Nile.

Thematic studies

The four geographical themes below should be investigated. These may be studied **separately, in combination with other themes,** or as part of the studies of places. Whichever approach is followed, these studies should be set within the context of **actual places** and some should use **topical** examples.

> "Taken together, the studies should involve work at a range of scales from local to national, and should be set in a range of contexts in different parts of the world. Contexts should include the United Kingdom and the European Union."
>
> DFE, *Geography in the National Curriculum* (HMSO, 1995)

You may study a theme in isolation, for example rivers, but you must use real rivers more than abstract worksheet ones. A river system in flood at the time of study can provide a motivating example for pupils who can use news videos, newspaper cuttings, as well as atlases, textbook or CD-ROM information.

Figure 2.8

You may study the most appropriate themes, for example PoS 8b and c, hotel building in your Caribbean locality PoS 4 and 5a, b and d.

The four geographical themes below should be investigated. These may be studied separately, in combination with other themes, or as **part of the studies of places.** Whichever approach is followed, these studies should be set within the context of actual places and some should use topical examples.

> "Taken together, the studies should involve work at a range of scales from **local to national**, and should be set in a range of contexts in different parts of the world. Contexts should include the United Kingdom and the European Union."
>
> DFE, *Geography in the National Curriculum* (HMSO, 1995)

Gives permission to develop themes relating to local areas, regions as defined by the teacher, countries or parts of countries.

Figure 2.9

The four geographical themes below should be investigated. These may be studied separately, in combination with other themes, or as part of the studies of places. Whichever approach is followed, these studies should be set within the context of actual places and some should use topical examples.

"Taken together, the studies should involve work at a range of scales from local to national, and should be set in a range of contexts in **different parts of the world.** Contexts should include the **United Kingdom and the European Union."**

DFE, *Geography in the National Curriculum* (HMSO, 1995)

Global or regions of the globe

Themes must be related at some stage to the United Kingdom and the European Union.

'Use appropriate vocabulary to describe and interpret their surroundings', and therefore schools need to define what is appropriate in their own context. (Figure 2.10 suggests a range of geographical vocabulary.) Other skills are very specific, for example, KS1 PoS 3c 'Follow directions including north, south, east and west' and KS2 PoS 3d 'Use the contents and index page of an atlas'.

Specific instructions are also given in the skills sections about the locational knowledge which is to be taught and three maps are included as part of the Order to specify this more clearly. This is in addition to the general need to develop locational knowledge.

Pupils' progress as a result of exposure to the programmes of study is to be matched to and assessed against a series of eight level descriptions included in the Order. These level descriptions form a 'broad brush' description of what might typically be expected of pupils working within a certain level. Pupils should be assessed against them at the end of a key stage and, for practical purposes, during a key stage. These levels, although they are imprecise and not able to sequence progression accurately, are generally an improvement on the myriad statements of attainment which were set out for assessment purposes in the original National Curriculum Geography (1991).

The 1995 Revised Order is in fact a passport to considerable freedom in primary geography. At first sight, especially for those who knew the 1991 Statutory Order, it is a minimalist curriculum. If some schools adopt its minimalist approach, especially in key stage 1, they could fulfil the letter of the law, but provide a limited geographical experience for their children. For many schools this could mean their practice would be restricted and the gains made through teaching the 1991 Order could be lost. However, this Order is a cleverly worded, although much reduced, version, as opposed to a radically changed version of the 1991 Order. A school could, therefore, easily take the 'maximum' approach and develop a very full geography curriculum from the 1995 Revised Order and, indeed, beneficially do more than required. Its very freedom is almost a shock to those who battled with the full and prescribed content of the 1991 Order.

Figure 2.10

Physical features	Climate weather	Settlement	Transport	Economic activity	Locational words
Key stage 1 hill stream slope river lake sea waves land soil rock pond steep gentle beach island valley mountain cliff wood forest waterfall	**Keystage 1** season desert wind rain cloud frost ice storm weather spring summer autumn winter storm weather	**Key stage 1** house shop park settlement village town city building site recreation ground library	**Key stage 1** road car pedestrian canal railway journey transport bridge tunnel zebra crossing traffic lights	**Keystage 1** shops work jobs farm factory service quarry mine woodland leisure centre hospital fire station health centre library	**Key stage 1** map plan country area place position north south east west near/far left/right up/down
Key stage 2 source meander tributary delta mouth erosion weathering deposition environment moon tides vegetation relief landscape features volcano earthquake	**Keystage 2** temperature rainfall mist fog dew precipitation reservoir sheltered exposed floods hurricane monsoon rainforest rain gauge anemometer	**Key stage 2** population urban rural density port resort hamlet pattern location housing estate suburb market centre conurbation	**Key stage 2** routes barrier network system atmospheric pollution subway	**Key stage 2** industry manufacture raw material labour fuel/power energy market natural resources supermarket business park tourism organic farming conservation pollution crop power station retail park acid rain pollution leisure	**Key stage 2** latitude longitude grid reference distribution globe region country continent county aerial

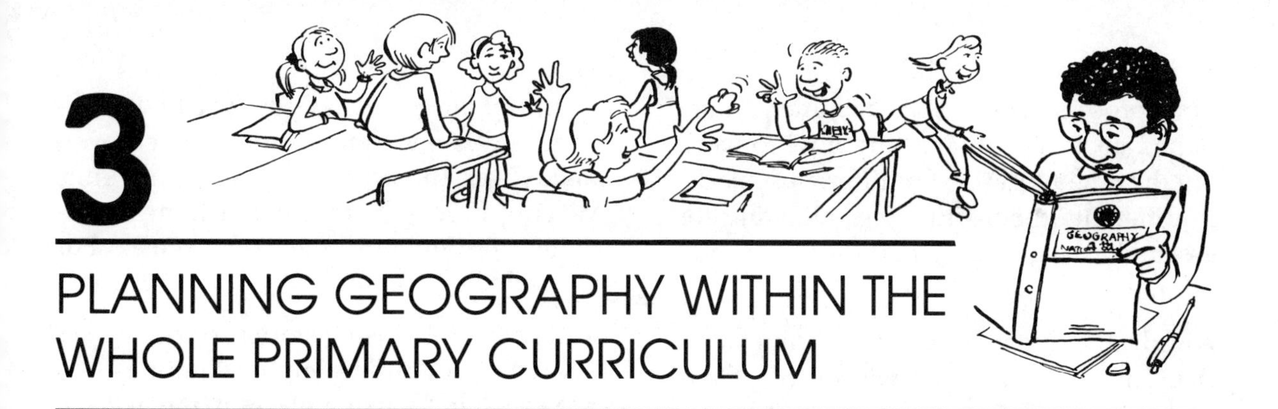

3 PLANNING GEOGRAPHY WITHIN THE WHOLE PRIMARY CURRICULUM

Where to start

With the introduction of the revised 1995 Curriculum Orders for core and foundation subjects, schools should be aware that a systematic approach to planning the whole primary curriculum is needed to give pupils their entitlement. So geography has to take its place with the other foundation subjects.

The Dearing review of the National Curriculum in 1994 identified the time management of the various subjects as a major problem with the implementation and delivery of the National Curriculum. So he gave clear time recommendations to the curriculum reviewing teams. This was to allow them to produce revised subject Orders with a manageable amount of content, when all the ten subjects and RE came together in schools. The teams then had to review the knowledge skills and understanding within each subject, and produce a curriculum to fit the notional time available.

The recommended time for geography and the other foundation subjects was:

- 36 hours in a school year in key stage 1
- 45 hours in a school year in key stage 2.

Another of Sir Ron Dearing's aims in this revision was to free up a notional 20 per cent of class time for schools to use in areas of the curriculum which they thought were particularly important in their own situation.

The School Curriculum and Assessment Authority (SCAA) then reinforced this with the publication of *Planning the Curriculum in Key Stages 1 and 2* in 1995, which was sent free into every school in England and Wales. This publication strongly recommends that schools adopt a whole-school approach in the planning and delivery of the whole curriculum and offers helpful guidance on how to do this.

Before you can start to plan your geography curriculum, whole-school decisions need to be taken about:

- The school's aims and objectives
- The length of the school day
- Time allocation for different subjects in each key stage
- What elements of the curriculum are to be included in the 20 per cent time allocation
- The structure of the planning for the core and foundation subjects
- The role of the curriculum co-ordinators in planning and supporting subject areas.

Many schools are in the process of reviewing and evaluating their existing curriculum plans and have already made many of these decisions. Others are still coming to terms with a whole-school curriculum planning process. The SCAA document mentioned above clearly outlines how to work out the time allocation for different subjects, in different school situations, allowing for assemblies, registration and the 20 per cent of time, etc.

An effective and motivating geography curriculum can be planned within the SCAA time guidance of 36 hours per annum (or 4½ per cent of curriculum time) in key stage 1 and 45 hours per annum (or 5 per cent of curriculum time) in key stage 2. This means that the authors do not assume that geography will spill over into the 20 per cent curriculum time. However, should a school decide that it is appropriate for the subject to do so, this can only be seen by the authors as beneficial.

Once decisions have been taken about broad time allocations over a year's teaching, then schools need to consider how this allocation will be divided up over the year. SCAA offers a model of *blocked, continuous* and *linked* units of work in and across curriculum subject areas. Many school have found planning this way to be successful.

A *blocked unit* can vary in length and is one which deals with only one subject area. For example in geography:

- Key stage 1 – a contrasting locality study
- Key stage 2 – an environmental issue.

A *linked unit* is where two or more subjects are planned from their subject frameworks but taught concurrently to give coherence in the classroom. For example:

- Key stage 1 – history and geography of the school building and school grounds
- Key stage 2 – science AT 3, PoS 2e (the water cycle) can be usefully linked with geography PoS 7a and b (the patterns and process of river formation).

A *continuing unit* is where specific knowledge, skills and understanding are developed over the whole key stage, in units varying in length from half a term, to ongoing throughout the key stage. For example in geography:

- Key stage 1 – using stories to widen children's geographical understanding
- Key stage 2 – using places in the news in a systematic way to develop children's place knowledge.

This very rigorous and subject-focused system has emerged in response to the need to provide a clear entitlement for pupils since the introduction of the National Curriculum in 1991. Because of the need to balance different subjects within the whole-school curriculum, schools have needed to move away from the most common planning tool pre-1991, the topic web. The most simple topic web model had a central starting point which was brainstormed with unlimited choice, and therefore had unpredictable outcomes. Even before 1991, HMI were arguing for greater rigour in planning:

> "*The weakest work in primary schools occurs when too many aspects of different subjects are roped together within integrated themes or topic work...generally topic work is difficult to manage; frequently lacks coherence at the initial planning stage and consequently is a fragmented experience for the pupils. The most serious casualties of this practice are history and geography.*"
>
> 'Standards in Education', 1989–9 Annual Report of the Chief HMI

Today topic webs, even where they are related to National Curriculum objectives, are still ineffective planning tools because:

- They cannot provide progression
- They cannot develop continuity across a key stage

- They do not assist monitoring and assessment
- They often do not emerge from or support schools' long-term planning.

To be able to deliver a broad and balanced curriculum, schools have to plan on different levels. Long-term plans across a key stage, worked in individual subjects but coming together as a coherent whole, lead into medium-term plans, i.e. lasting usually for half or a whole term. Short-term planning, usually spanning a week or a day, further breaks down the curriculum content to support the teaching and learning in the classroom. Currently there is a continuum of practice nationally which ranges from the topic web with National Curriculum objectives, through to schools that have planned using the SCAA model, to schools which have developed their own planning model that takes account of a variety of different planning strategies which have evolved over time from a topic planning model.

We are not saying that integrated subject planning, often referred to as topic planning, cannot deliver the National Curriculum. Many schools, especially when planning key stage 1, use multi-subject focused topics effectively. However if this is the case, long-term planning must ensure that the relevant PoS aspects are rigorously distributed across the topic titles and adhered to. It is not the choice of topic title that is important, but the mixture of PoS that is the key to an effective learning programme. It is better to start by clustering different PoS from a range of subjects, than by choosing a title and seeing what fits. This type of curriculum planning needs to be monitored closely to ensure progression and avoid repetition. It is still easier to maintain multi-subject planning within key stage 1 than towards the upper end of key stage 2 due to the increasingly specialised nature of the PoS for this age range. The older the pupil, the greater the need for a topic to focus on fewer or one subject at a time in order to develop the learning objectives for the various subjects effectively. OFSTED supports this view – evidence enabled it to state that:

> *"Most primary schools taught geography as part of broader topics, but in Years 5 and 6 the sharpening of the geographical focus at times and identification of clear learning objectives improved standards achieved."*
>
> OFSTED, *A Review of Inspection Findings 1993–4* (HMSO, 1995)

Planning needs to be equally rigorous in both key stages but the planning focus of both content and concepts of subjects will need to be more rigorous to ensure progression. Schools will need to consider units of work of differing lengths. Two, four or six weeks, half-term or even a term duration may be appropriate. In key stage 2 several units of work may run alongside each other over the course of a year, depending on their focus as shown in Figure 3.1.

Whether the SCAA model of blocked, continuous and linked units of work is used for geography, or whether some rigorous form of topic planning is the method used in your school, you still need to be aware of the appropriate links of geography with other subjects. (See Figure 3.2.)

Linking with other subjects

Geography is but one element of the whole curriculum. It would not be helpful for this book to imply that geography should be the centre of the curriculum. Instead, it seeks to explain the principles of geographical teaching and learning so that teachers can work with these and build geography into the whole curriculum. It is also recognised that

Figure 3.1

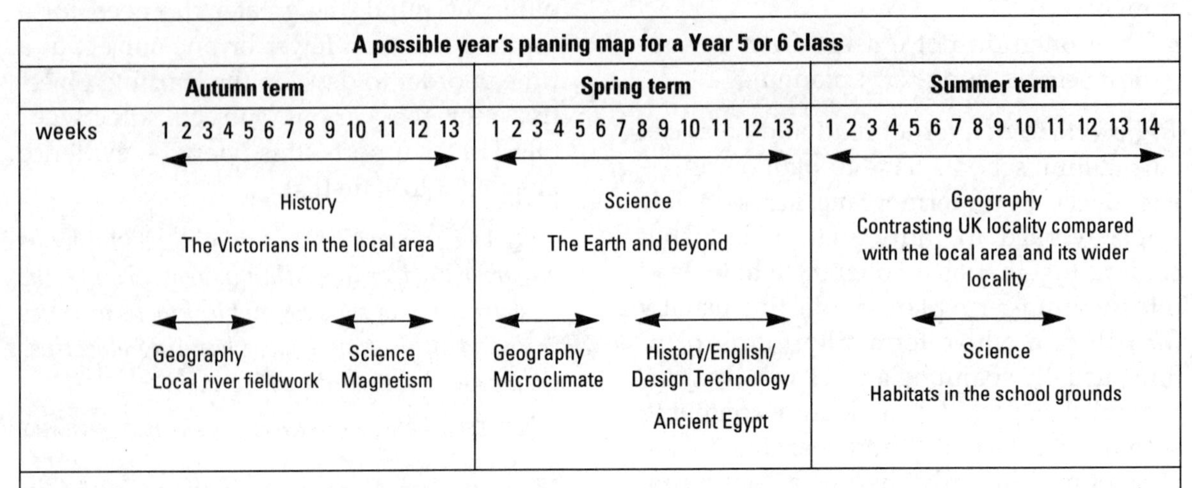

Figure 3.2

Where planning is weak	Where planning is strong
There is no overview of all subjects in any one half or full term. The key stage plan does not cover all aspects of all subjects required. No decisions have been made about time allocations, across the year or for individual units. The work in each half-term is not prioritised so it fits within the time allocation. The science curriculum is the leading subject that controls topic plans.	Long-, medium- and short-term plans exist These are realistic in terms of content/time and planned directly from the different subject PoS. Decisions have been taken about the balance of subjects across a year.

there will be occasions when teachers may wish to deal solely with geographical work. The older the children, the more likely it is that this kind of focus will become sensible.

Before tackling the planning of key stages and breaking these down into units of study, we need first to understand exactly how geography links with other core and foundation subjects in order to make planning more manageable and learning more motivating. These links are outlined in the following section of this chapter. Links with cross-curricular themes can follow, and these are referred to in Chapter 12.

How can we work with science?

Geography is itself both a science and an arts subject, and is often described as the bridge between the two. Nevertheless, until the advent of the National Curriculum primary schools had tended not to exploit the strong links within this subject area. The original science and geography Orders overlapped in many areas. This overlapping tended to confuse teachers and cause unnecessary duplication. Common sense beyond the National Curriculum tells us that science and geography have many overlaps as shown in Figure 3.3.

Figure 3.3

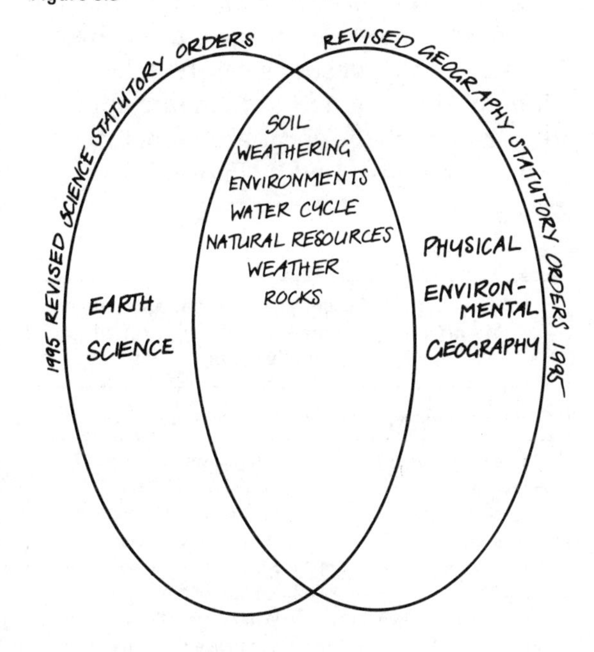

However, at primary school level, for the purposes of clarity, the Revised Orders for science and geography eliminate overlap. Weather is now firmly part of geography, see KS1 PoS 5c, KS2 PoS 8a, b, c and d. Soils and rocks are now to be found in science AT3 KS2 PoS 1a, KS2 PoS 2a and b as part of Materials and their properties. The water cycle similarly appears here, under PoS 2e (and linked to c and d). Weathering is now an explicit part of the primary curriculum. Weathering and rock types are dealt with in depth at key stage 3 of the science curriculum.

Nevertheless, although this clarity is helpful for NC planning purposes in some ways, we must remember that real environments do not divide themselves neatly into subjects, and a little common sense is needed to give children coherent learning. For example, if in geography your school is in a coastal locality, as part of studying your local physical features, you would surely visit the beach and talk about any other natural features such as cliffs. Many teachers would name the type of rock, especially if it is a well-known type, such as chalk, sandstone or granite. This arises naturally as part of geography, whether or not samples of local rocks and soils are to be collected and taken back to school for science AT3 work.

If a school wishes to include fieldwork on weathering as part of local environmental work in the context of science or geography, or indeed both, then as long as it can be seen to be teaching all the other explicit parts of the programmes of study for those subjects, and it has sufficient time to do so, there is nothing to stop it doing so.

Many teachers are likely to make links between the water cycle in science and weather and rivers in geography, either by planning a dual subject-focused topic in a key stage 2 plan, or by linking together two blocked units on the water and rivers or weather within a whole-school curriculum plan.

One word of caution when linking science and geography, however: as pointed out in Chapter 1, the content of geography relates to physical and human environments, and the interaction between them. We must take care to emphasise the effect of humans on scientific processes in the environment, otherwise it is science that is being taught, not geography and science.

The process of geographical enquiry is very similar to that of scientific investigation as required in Sc1. Thus active learning in both subjects is mutually reinforcing. It is important to recognise that the science/geography link is far stronger than the history/geography one!

Don't forget history partners!

Traditionally history has been the partner for geography. Links between the two have been very strong in the form of integrated humanities topics. However, sometimes the teacher's rationale for the history/geography link can be very tenuous and has resulted in a lack of clarity in the concepts which apply to the respective subject areas.

The history Order is more concept-led than the geography Order, although the history programmes of study specify content quite clearly. Because the history study units for key stage 2 and the contents of the geography order are both so prescribed, many primary schools no longer consider it realistic to link the two subjects at all. This is regrettable, as there are some areas where there are clear links which can produce excellent work so long as the concepts and skills relevant to the discrete programmes of study are recognised. Beware the incidental or spurious link: is a great deal of map drawing related to the history study unit on 'Romans, Anglo-Saxons and Vikings in Britain' really the best context for developing mapwork concepts? It may be a useful opportunity to practise or apply skills.

There is, however, one major area – local area work – where history and geography Statutory Orders do genuinely overlap, as Figure 3.4 illustrates.

Figure 3.4

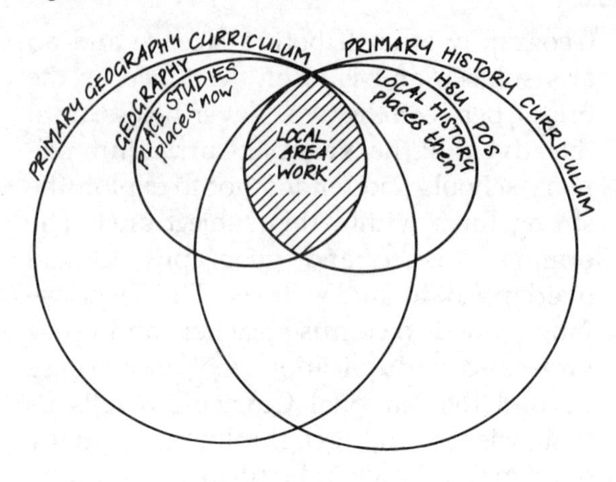

Chapter 1 outlined the importance of the use of key questions in geography work. The same key questions can be asked in a different time context to motivate children to learn about the local area, present and past, as demonstrated by Figure 3.5.

Figure 3.5

Key questions for geography local area and history study unit on local village/area of town/whole town	
Geography (now)	**History (then)**
Where is this place? What is it like now? Why is it like this? How is it changing? How can we find out about it now? What is it like to live there now?	What is/was this place? What was it like in...? Why was it like this? Why did it change? How can we find out what it was like then? What would it have been like to live there then?

It is important to notice the differing use of 'local' in the geography Order and the history Order. The 'local area' is clearly defined in geography for key stages 1 and 2 (see Chapter 10). It is not defined in the history programmes of study, so can be as large or small as teachers wish, but they need to be clear how this affects any linked geography teaching.

These key questions should naturally be applied to the model for the enquiry process noted in Chapter 1. The same process can be applied to investigations in history so that it is possible to collect data on past and present situations in the local area. This process enables the concepts of similarity and difference between now and then to be developed, as well as that of continuity and change in history.

The concept of change in history and geography can sometimes become confused. When teaching geography, teachers can go into such historical detail about change in a place that they are actually teaching historical geography or pure history. Teaching settlement patterns in Roman Britain is *not* within the remit of geography in the National Curriculum. A brief reference to the fact that a locality grew up in Roman times because of its geographical location and site is. Such a detail may be fundamental to a geographical understanding of the physical and human features of the town now.

Geographical change in the context of the Order is recent or proposed change (i.e. change which is likely to happen in the future). Recent change in geography, when referred to or studied in any depth, needs to be relatively recent. Depending on its context, it may go back to last year, for example, 'Is there any less traffic coming through the town since the by-pass was opened?' (assuming data was collected to demonstrate the difference). It may go back several decades, for example a class may use modern aerial photographs alongside recent and 25-year-old OS maps to carry out an enquiry on how their settlement has grown since their parents were born. Much geographical change work can arise about proposed change. Environmental issues and development issues in settlements are ideal opportunities to explore through mapping and role play – 'What would happen, if...?'

Some of the history study units in key stage 2 do lend themselves to a link with geography.

For example, a topic on Egypt could have a two-subject focus: 'Ancient Egypt' (history study unit) and a study of a locality in present-day Egypt (geography – a locality in Africa), which offers opportunities to identify similarities and differences and to compare past with present. This is better tackled with older juniors as younger ones can confuse ancient and modern Egyptians.

Most history study units are best kept quite separate from geographical learning. However, at key stage 1 it is easier to adopt a multi-subject-focus topic which can deal clearly with the history and geography Order requirements. For example, in a topic on 'Homes', pupils can start by talking about and drawing their homes today using geographical vocabulary, before they go on to ask questions such as, 'Is my house old or new?' or 'Have machines in my kitchen always been like this?' Similarly, an infant topic on transport may begin with how pupils travel to school, using graphs drawn from pupil data. Then pupils may draw a map of their mind's-eye route to school before going on to ask how their parents and grandparents travelled to school. This key question approach is fundamental to both Orders.

Mathematics and geography go hand in hand

Mathematics is indispensable to geography as a tool in the enquiry process. Good practice in both subjects enables pupils to work on their own, real-life data. Maths is essential in geography to:

- Collect data
- Record it
- Present it
- Interpret and analyse it.

The following examples relating to data collection in fieldwork and secondary source work should illustrate the point.

1 Fieldwork In order to understand how site conditions can influence the weather and how the weather varies over place (KS2 PoS 7) and to use instruments to make appropriate measurements (KS2 PoS 3b), it is necessary to collect weather data in different sites in the school grounds. This could be done by simply relating temperature and rainfall. Numerical data collected could be represented as a bar chart (rainfall) and a line graph (temperature). The data represented for the various sites needs to be compared in order for the graphs to be analysed and judgements to be made about the warmest/coldest/ wettest/ driest areas of the school site.

2 Secondary source work When using information in, for example, books, leaflets or posters on a locality in an economically developing country, an older pupil may need to sift through figures and statistics or a series of pie charts in order to answer the question 'How do people in this Indian village spend their day?' Having found such data, they then need to interpret it and present it in their own way, for example in percentage figures or as a summary in words.

The whole area of graphicacy skills in geography draws on maths. Indeed some of the examples in the mathematics Statutory Order come straight from a geographical context, for example map references. Many concepts are shared by the two subjects at primary level:

- Distance
- Direction
- Area
- Links
- Networks (patterns)
- Shape
- Size
- Scale
- Time.

For older juniors, communicating facts about weather, climate, farming, industry and population through mathematical units and quantities lays the foundation for the description and analysis of these areas of content at the later secondary school level through the use of more complex statistics.

The maths ATs which need to be closely examined when planning geographical work are: Ma2, Shape and space and measures; Ma4, Data handling; Ma1, Using and applying mathematics. Maths is a key skill in the whole curriculum and evidence of number work will be looked for in geography as part of OFSTED inspections.

HMI, in their report on the first year of National Curriculum mathematics, 1989–90, said that the AT concerned with using and applying maths received little attention. The situation has only slightly improved since then. Geography can provide splendid opportunities to develop these maths skills. An analysis of maths schemes and textbooks will show a wealth of geographical maths exercises in isolation, as in the maths Statutory Order, for example traffic surveys, distance measuring, plan drawing, graph drawing. Sometimes it is appropriate to set pupils these isolated second-hand exercises for practice and reinforcement, but more often than not it can make sense to rationalise the maths scheme. Do this by omitting exercises which could be done in a geography enquiry context by using and applying maths within topic work. A common example is the over-used traffic survey. Pupils could predict the volume of traffic types they expect to pass the school within a certain period of time, then collect data in tally chart form, graph the results and compare their predictions. The geographical context for this exercise could be work on human geography and environmental issues in their local area.

As part of their school policy and in order to save time in an overloaded curriculum, maths and geography coordinators should:

- Analyse the school's maths scheme for geographical activities or exercises
- See how these might be better replaced or applied directly to first-hand data collection in geographical, scientific or historical work
- Take decisions about which exercises or activities to keep, omit or use as a reinforcement option
- Take decisions about classroom management; for example, will some groups use the maths scheme activity rather than the real-life data in order to free the teacher to help the group working on their 'real' data? Consider how to ensure that all children have equal access to the different experiences
- Make sure that pupils and parents are aware that important maths work is being done, even though it may not be 'in the maths book'.

Geography as a medium for English

Developing the use of language is a requirement common to all National Curriculum subjects since the introduction of the revised subject Orders, and it is clearly specified in the preamble to the geography programmes of study for key stages 1 and 2:

> "*Pupils should be taught to express themselves clearly in both speech and writing, and to develop their reading skills. They should be taught to use grammatically correct sentences and to spell and to punctuate accurately in order to communicate effectively in written English.*"
>
> DFE, *Geography in the National Curriculum* (HMSO, 1995)

Reading, writing, speaking and listening are key skills to be practised through geography, and the OFSTED inspection framework inspects the subject for its contribution in these areas.

It is not possible to teach geography without using and improving core English skills. It would indeed be possible, although not desirable, to teach English at primary level solely in a geographical context using both language and literature. Listening, speaking, reading and writing are central to English and find a natural context in geography so long as we are aware of the requirements of both.

As with maths, a greater recognition of how we can develop English and geography simultaneously will save valuable teaching time and provide a more meaningful learning context for children. Figure 3.6 summarises how geographical activities help to develop work for the attainment targets in English.

Considerable emphasis will be placed on speaking and listening in geographical topics with infants, whereas it will tend to swing to reading and writing skills as the pupils grow older. Upper juniors should be dealing with higher order reading and writing skills using geographical texts *Developing Tray* is a useful computer programme for developing English in a geographical context. Again, staff, parents and pupils need to be aware that English in a geography context is a valid medium for good practice in English teaching and can fulfil the English Statutory Order. Teachers can highlight this by:

Figure 3.6

Developing English through geography

En1 Speaking and listening

Speaking: describing photographs, describing postcards, talking about maps, using geographical vocabulary, giving instructions or directions, interviewing during fieldwork, role play to present an argument, answering geographical questions, asking/answering open-ended questions in a whole-class group, discussing in pairs or small groups, explaining geographical ideas.

Listening: teacher or peers reading a story based in a particular location or locations, factual information on television, radio programme, tape, etc. listening to a variety of speakers – peers, adults, in groupwork – listening to geographical ideas.

En2 Reading

Information from travel literature (advertisements, brochures, etc.), planning outlines, descriptions of places, descriptions of processes, descriptions of weather, encyclopedia and other classroom or library reference books, reading scheme books which are located in a particular place.

En3 Writing

Factual recall writing – descriptive, summarising, paraphrasing; creative writing – a story with a place base after studying a particular locality; analytical writing – explaining information gathered in fieldwork, note taking in fieldwork, note taking during television or radio programmes.

The use of geographical vocabulary where appropriate – focusing on the spelling and meaning of words currently being used connected with geography work, e.g. 'beach', 'cliff'.

Geographical words and passages can be used to practise handwriting.

- Explanation of the skills being developed
- Displays of work which pupils and/or teachers can label for subject-linked concepts and skills.

A word of warning about reading ages: most geographical textbooks and information books written prior to 1991 have a reading level suitable for literate, articulate juniors. The majority of school libraries contain many such books still. A resource deficit has existed here, but as we go towards the millennium, authors and publishers are at last addressing this reading age issue. A much wider range of reading-age appropriate books suitable to support geography now exist.

Geography and design technology

The Revised Design Technology Order states that children should develop their capability in design and technology in three main ways:

- Through designing and making activities (DMA)
- Through focused practical tasks (FPT)
- Through investigation, disassembly and evaluation of different products (IDEA).

There are few links between the FPTs, IDEAs and geography, but there are some links between DMAs and geography. It is important, however, to recognise that when there seem to be links, the learning objectives need to be made explicit in each National Curriculum area. Also, the time allocation between the two subjects needs to be clear.

In many schools the weather aspects of the geography Order are used as an opportunity for children to design their own methods and equipment for recording weather measurements. Simple anemometers, weather vanes and barometers can be designed and used. Commercial ones should be used alongside or after these, or for evaluative purposes. We need to be aware of the criteria needed for such instruments to perform the task for which they are designed with reasonable accuracy. There are a variety of pupils' and teachers' books relating to the weather to help here. It must be remembered that the designing and making of the instruments is technology and the using of them is geography. So the appropriate medium- and short-term planning, plus the time allocations for the two subjects in the week it is taught, must reflect this.

Many of the enquiry questions and focused statements in Chapter 1, Figure 1.2, could be used to develop the concepts of both subjects areas, for example 'How can we improve our school grounds?' and 'Where is the best place to site the new flower beds we are going to design in our school grounds?' In designing parks, improving playgrounds and adventure playgrounds, children have to consider the environmental impact of their designs, so the work can be closely linked to KS1 PoS 6 and KS2 PoS 10 (Environmental geography). Geography can support the planning phase of many technology tasks by enabling the children to use their mapping skills, as a tool to support their planning of the park, adventure playground, etc. on a waste ground site. These types of activity further support environmental understanding and their local knowledge will be used in choosing a site.

Thought can be given in technology to developing alternative technologies to support the life style of the people being studied in geography who live in an economically developing country. In this context technology IDEAS work could also be included.

Geography helps children apply what they are learning in design technology:

- Stressing the importance of technological developments in encouraging economic activity, explaining patterns in economic development
- Stressing that alternative technology is often more appropriate than grandiose solutions in developing localities; for example, simple mechanical irrigation systems may damage the environment less and keep more villagers employed than a hydro-electric power scheme with an electronically-controlled irrigation scheme.

IT – the perfect tool

Information Technology (IT) has the status of a core skill in the Revised Curriculum Order. All pupils in key stages 1 and 2 have an entitlement to develop IT skills and capabilities. The Order states:

> *"Pupils should be given opportunities, where appropriate, to develop and apply their IT capability in their study of National Curriculum subjects."*
>
> DFE, *Information Technology in the National Curriculum* (HMSO, 1995)

So teachers have to develop their pupils' IT capability within their studies of the other National Curriculum subjects. For geography, this is strengthened by specific references to IT in the geography Order. The Common Requirements state:

> *"Information Technology: Pupils should be given opportunities, where appropriate, to develop and apply their information technology (IT) capability in their study of geography"*

in KS1 PoS 3f:

> *"Pupils should be taught to: use secondary sources e.g. pictures, photographs (including*

aerial photographs), books, videos, CD-ROM encyclopaedia, to obtain information."

and in KS2 PoS 3f:

"use IT to gain access to additional information sources and to assist in handling, classifying and presenting evidence, e.g. recording fieldwork evidence on spreadsheets, using newspaper on CD-ROM, using word-processing and mapwork packages."

DFE, *Geography in the National Curriculum* (HMSO, 1995)

The use of IT encourages the development of the problem-solving or enquiry approach to learning which is pertinent to geographical enquiry. It enables children to explore abstract ideas and test theories. It encourages group work and collaboration between pupils. IT can also help in the presentation of children's work, leaving them to concentrate on the content. As with all skills, IT should not be practised in isolation but in real situations. Geography provides a perfect curriculum area in which to work.

There are two main strands within the IT PoS:

- 'Communicating and handling information' in both key stages 1 and 2
- 'Controlling and modelling' in key stage 1, which develops into 'controlling, monitoring and modelling' in key stage 2.

Communicating and handling information Communicating information can be done in the form of pictures, words and numbers. It can be used to develop or revise ideas. It enables pupils to communicate with a wide range of audiences using word-processing packages, desktop publishing, electronic mail systems, and graphic packages. All these types of programs can be used to good effect in geographical work.

Handling information deals with the storage and retrieval, modification and presentation of data. It examines patterns and relationships in information and teaches pupils to access information from a variety of sources. It uses database packages, spreadsheets, CD-ROMs and remote data bases. Geographical enquiry provides relevant data for pupils to work with, enabling them to store and manipulate information they have collected themselves. If IT is used to process and display information collected by the pupils, more time will be available to question the data and test hypotheses.

Controlling and modelling This involves using computers and electronic devices to control and simulate environments. It uses data-logging devices, programmable toys, concept key boards and control software. Programmable toys such as the Roamer help children to develop spatial awareness and practise their mapping and directional skills. Software is used to create real and imaginary situations, adventure programs, spreadsheet and programming languages such as Logo. These can be used to investigate environments that cannot be investigated at first hand. Using map adventure programs and directional language develops spatial awareness, mapping and directional work. Modelling can help children to recognise patterns and relationships between different sets of geographical data.

Monitoring Data loggers and sensors can be used on their own or connected to computers to monitor and store geographical information. Sensors can be used to collect data, for example weather data, humidity, length of sunshine or soil temperatures for analysis. Computers can be used to sense and record these changes in the environment over long periods of time and without large amounts of data being entered by hand.

Because of the flexibility of IT the opportunities to use it in geography are endless. The Geographical Association (GA) and the National Council for Educational Technology (NCET) have produced an extremely useful leaflet called *Primary Geography: A Pupil's Entitlement to IT* (1995). This can be obtained from either organisation (for their addresses see Chapter 13, Resources). This leaflet clearly outlines how IT can support and enhance pupils' geographical understanding, while providing a suitable vehicle to develop their IT skills. Figure 3.7, reproduced with permission from the GA and NCET, clearly demonstrates this.

Art from other places

Sensible links can be made with the art programmes of study and geography. Both Revised Art Order PoS require pupils to be introduced to the works of artists, craftspeople and designers from their own and other cultures, Western and non-Western. The images of distant localities and their artefacts can support pupils' learning in geography about the distant place.

Music from other places

Similarly, it is possible to develop linked geography and music units of work. The Revised Music Order requires pupils to perform and listen to music from different places and cultures. Tapes of music from European countries, India, Egypt, the Amazon area and various other locations around the world are now fairly readily available in the UK. Some geography distant place packs include taped extracts of music, capitalising on the link between music and distant place geography work. Pupils from Muslim or Caribbean cultures may be able to bring real examples of their musical heritage to geography work if appropriate to the place.

Geography and orienteering with PE

Much infant geography is intrinsically linked to spatial awareness work in PE through the Games and Gymnastic programmes of study in key stage 1. Teachers use these as a context for geography directional work in a practical way before moving to mapping work or cardinal points. In key stage 2, the programme of study for PE similarly emphasises direction changes, but adds another area, PE PoS 5 (Outdoor and adventurous activities) where a useful linked unit of work developing geography skills and PE orienteering could be devised. Orienteering can require the use of some or all of these geographical skills: recognition of physical or human features, sequencing features, use of maps at a range of scales and compass use.

Auditing current practice

Once you have a clear idea of the demands of the revised 1995 National Curriculum geography and its links with other subjects, you can begin to plan how to deliver it in your school. In order to do this effectively you need to audit the current state of geography teaching and learning in school.

Decide first who will carry out the planning for the key stages. It could be:

- The headteacher and/or deputy
- The whole staff (especially in rural schools)
- A team of teachers
- A subject coordinator
- A coordinator and advisory teacher, or other outside consultant together.

Here it will be assumed that the planner is the geography coordinator.

Figure 3.7

How can IT help pupils' learning in geography?	**When undertaking geographical activities pupils will:**	**IT can enhance this activity by:**
by enhancing their skills of geographical enquiry	• ask geographical questions • observe, record and investigate data from fieldwork and secondary sources • create, use and interpret maps at a variety of scales • communicate and present findings	The use of • databases, spreadsheets, or data logging equipment, e.g. for a study of shopping, farming, or the weather • software to present information in a variety of ways, e.g. text, graphs and pictures • concept keyboards to investigate images of localities and develop map skills
by providing a range of information sources to enhance their geographical knowledge	• draw on appropriate sources to obtain information, ideas and stimuli relating to places and geographical themes • become familiar with and use geographical vocabulary	Providing access to: • people and first-hand data using Electronic mail (e-mail) and fax • photographs, video sound and other information, e.g. on CD-ROM, to study another locality or environment.
by supporting the development of their understanding of geographical patterns and relationships	• recognise patterns, and make comparisons between places and events	The use of: • databases and spreadsheets, simulations and multimedia to provide an insight into geographical relationships , e.g. weather patterns, changes in traffic flow or causes and effects of water pollution • a floor turtle to develop spatial awareness
by providing access to images of people, places and environments	• develop an awareness and knowledge of the culture and character of places	Providing access to: • people and first hand data using e-mail and fax • TV photographs, video, sound and other information, e.g. on CD-ROM, to study another locality or environment.
by contributing to pupils' awareness of the impact of IT on the changing world	• use specific examples to illustrate how IT influences communication, leisure and the world of work	Creating opportunities to discuss how computers are used to: • book a holiday • control stock in supermarkets • transmit information via satellite communications • forecast the weather

From Geographical Association and the National Council for Information Technology, *Primary Geography: A Pupil's Entitlement to IT* (NCET), 1995

Auditing involves the coordinator, head teacher and staff taking stock of:

- What geography is currently being done in which years (even though it may not be being taught under a geography label)
- Which geography skills are covered in which years
- What resources for geography already exist in school
- What methods of assessment and recording exist.

It has been said, quite understandably, that primary teachers are 'sick to death of auditing'. Yet it remains a useful activity which will enable you professionally to:

- Assess current practice
- Provide a way forward in the planning process
- Actually save time in the long run.

Auditing part of the current curriculum is rather like taking on a new class. A good teacher will want and expect to know what pupils have done before and what they should do next, so that continuity and progression can be maintained. In order to promote continuity and progression in learning geography, the same applies.

Guidelines for auditing

1 Use the photocopiable programme of study aspects tables (A1 and A2 pages 205 and 206, and see Chapter 2) as an audit of what geographical learning is already taking place. The tables can be used in two ways:

a) Pass the appropriate key stage table to each member of staff in turn and ask them to examine their year's topics for geography or their year's specific geography teaching. They should tick off PoS aspects that they have worked towards with their pupils/groups of pupils. (In vertically grouped classes the teacher may prefer a separate grid for each age group.) Each member of staff should use a different colour or symbol to record. Alternatively copy the grid onto overhead projector transparencies and give one to each member of staff to tick as appropriate. Then overlay them and project to get an idea of strengths and weaknesses in any particular aspects and in any particular year. Yawning gaps will show up, as will areas of constant repetition and lack of progression. This exercise can then act as a guide to planning: the visual summary provided by the tables will indicate to staff and coordinators where they need to reassess and plan very carefully.

b) The same process can be undertaken by individual teachers on a single unit they have taught to establish how its content fulfils the Revised Statutory Order at their pupils' relevant levels. The completed grid can then be used as a prompt to establish which parts of the programme of study need to be planning for if the topic or unit of work is repeated.

2 Ask all staff to list current geography resources against the checklist in Appendices A3 and A4 (compiled from the evidence and implications of the programme of study of the Revised Order and the authors' wide experience of working in and with schools on inspection or consultancy). Decide which are lacking, prioritise them in order of need, balanced against budget availability.

3 Audit current record keeping and assessment systems for geographical work. (According to HMI and OFSTED inspectors, it is not unusual to find that they do not exist!) In order to ensure continuity and progression as well as to comply with legal teacher assessment requirements, they will need to be designed (see

Chapter 6) if they do not already exist.

4 Examine sensitively with staff how much the key questions and the enquiry process are being used in geographical work to date. The answer could be 'not at all' – so bear them in mind during in-service work and when planning units of work (see Chapter 5).

5 Decide on the types of planning approach through which geography will be taught.

Having undertaken this exercise, you are now in a position to plan to meet the needs of Revised National Curriculum geography.

4

PLANNING YOUR GEOGRAPHY CURRICULUM – KEY STAGE PLANS

Why plan a key stage?

Key stage planning has, over the last five years, become the recognised first step in planning a coherent curriculum to support the National Curriculum Orders. Planning needs to be looked at on the macro scale to avoid repetition, aid progression and continuity in the child's geographical experience. Teachers need to look closely at the links between geography and other subjects. By establishing the connections and supporting experiences between the content of different subjects, it is possible to plan for the children a coherent learning experience, where the various bodies of knowledge are neither fragmented, nor linked in a forced manner.

In the 1995 Revised Geography Order the amount of prescribed knowledge, skills and understanding has been significantly reduced. This makes the subject more manageable. However an additional problem now arises – many of the individual programme of study paragraphs, although few in number, contain within them a great deal of geography which needs to be unpacked. These paragraphs also presume an understanding of the depth and breadth of the subject. Schools have responded to this in a variety of ways, some adopting a minimalistic planning approach, others developing a maximalist approach.

The Revised Order still offers schools the flexibility to plan their geography curriculum to suit their individual circumstances, building on current practice, links and any particular strengths you or your colleagues have. When dealing with long-term key stage planning, you need to consider the possibility of using a variety of different length units of work (see Chapter 3). There is no ruling on how long a unit of work must be, and a flexible approach to this helps to balance the topic or project element of the 'whole primary curriculum'. Schools must decide this for themselves with reference to the SCCA time guidelines. Obviously an in-depth study of the school locality would take a term, but a look at a river in the locality could be covered in four weeks, and the other aspects of the locality taught at another time. In all planning the authors have been involved in, a mixture of unit lengths has been found to be best. Consider units of any length between two and twelve weeks. Also, a mixed approach where subjects are taught separately and in a linked way seem to produce the best results. A balanced approach to all National Curriculum subjects is needed to give pupils their full entitlement curriculum. Flexibility helps to create a broad balanced primary curriculum.

The places, themes and skills in the Order were not designed to be taught in isolation.

- All themes need to be taught in a place.
- All place studies involve themes.
- All place and thematic studies need geography skills to develop.
- **Therefore, geography skills should not be taught in isolation.**

Figure 4.1 based on the cube model illustrates this.

Figure 4.1

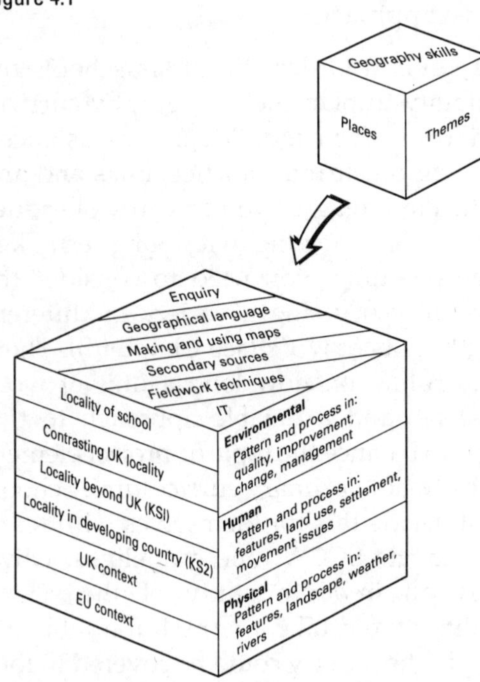

A long-term or key stage plan needs to be developed, piecing together and using up all the places, skills and themes. Units of work, or medium-term plans, need to be developed from these. This represents a full scheme of work for geography. One aspect of planning cannot exist on its own; both elements are needed to support the teaching and learning of geography in the classroom. The key stage plan element of the scheme of work can be individual to geography or documented alongside other subjects. This shows the *coverage* element of the planning for the scheme of work. The medium-term, or unit of work, planning side of the scheme of work shows *how* these PoS will be taught.

> *"Many schools do not have a scheme of work for geography which can be used to ensure continuity, consistency and progression. Too many make do with a generalised policy document on geography..."*

OFSTED, *Geography: A Review of Inspection Findings 1993–4* (HMSO, 1995)

This main finding from the above document reinforces the authors' long-held viewpoint and experience in the importance of both aspects of planning being there, to support teachers in providing a quality geography curriculum. At some point in their planning process, schools must look at the key stage plans for all National Curriculum subjects and get the correct balance of work across the years and the key stages.

The units in the key stage plan should provide a balance and breadth of geographical experience to the children, while allowing for progression, differentiation, different teaching and learning styles, school situations and pupil experiences, difficult as that may be.

Key stage planning helps schools to address these major issues:

Curricular issues

- Progression within the key stage
- Links with other subjects
- Balance of places, themes, skills and issues within and between years
- Enquiry process
- Opportunities for fieldwork
- Resources
- Continuity between KS1, KS2 and KS3
- Cross-curricular themes, dimensions and skills.

School issues
- Time allocation
- Blocked, linked and or continuing units of work
- Subject-focused topics, blocked, linked or continuous
- Pupil groupings, horizontal or vertical
- Staff expertise and commitments
- Policy for planning, long term and short term
- Subject budget
- Cross-phase liaison.

Where are you?

There is more than one way to design a key stage plan. Establish where your school is now, what is happening in the primary geography curriculum. Schools will be in different situations depending on the amount of work they have done on planning the previous National Curriculumn Order. In some schools the 1995 revisions have provided the opportunity to review and evaluate the whole primary curriculum, changing the way many subjects are planned and taught. Your school could be:

- Evaluating and revising a current key stage plan that is working well and this planning exercise will enable you to look more closely at progression and continuity
- Evaluating an existing plan with the opportunity for improvements
- Reviewing the whole-school curriculum planning
- Just beginning to address geography through the Revised Order

or at any point between these examples. Take a close look at where you are and review the position, evaluate the successes, the resources and current practice. Are there elements of that practice you want to keep? This might mean using some of your 20 per cent of freed time; it might not.

How to start

Planning a key stage is a logical process (see Figure 4.2). Before you begin you need to have established, as a school, how you are going to manage your curriculum time. As mentioned in Chapter 3, the SCAA booklet, *Planning the Curriculum at Key Stages 1 and 2*, is useful here. As a rough guide, geography in line with other foundation subjects has a recommended time allocation of:

- Key stage 1: 4 ½ per cent or 36 hours
- Key stage 2: 5 per cent or 45 hours.

This needs to be blocked into different sized units of working according to the way your school has decided to work. Some schools are following the SCAA guidelines of blocked, linked and continuing units. Others plan some geography in every term, others divide the six half term units equally between geography, history and design technology. As with all National Curriculum subjects it is a pupil's entitlement to cover all of the PoS.

Having established aims, objectives and audited current practice, parts of the PoS can be clustered together and the units they make can be spread across a matrix, following the style your school has chosen to follow.

The most efficient way to make a key stage plan is to distribute the localities which need to be taught across the key stage, consider the themes and then finally the skills. The minimum entitlement would be to deal with each locality and place context (wider context, United Kingdom and European Union context) once, but this would provide a poor geography curriculum.

For key stage 1, good practice in geography splits the local area into:

- Classroom and school buildings
- School grounds
- Area around the school.

Geography embracing all these areas needs to be planned for, as well as the contrasting locality elsewhere.

In key stage 2 the local area may still include the school grounds and/or building if a school deems it appropriate. For example, KS2 PoS 8a 'how site conditions can influence the weather', micro climate work is best introduced practically in the school grounds. In addition it may be appropriate to have a continuous unit of work in one year taking weather measurements in the school grounds and logging them (good opportunity for IT here!) over two or three months to develop work on KS2 PoS 8b about seasonal weather patterns. The local area outside the school, both close to it and further away, PoS 4 and 5e, 1d, need to be taken account of. So to build in progression about the geography of the local area, it makes sense to include some entitlement to it in every year. Just how much and how it is dealt with will be an individual school's decision, based on their local resources. See Chapter 10, especially Figure 10.3 (page 125) for auditing local possibilities and the section entitled 'And finally – your contrasting locality queries answered!' (page 136) for full details explaining locality and place contexts.

Building up children's understanding of the various physical, human and environmental aspects of the local area is a key to helping them to understand distant places – for them to make comparisons between local and distant, through identifying similarities and differences – so they need continuity to develop a depth of understanding about their local area and the themes within it.

In key stage 1, there is a need to cover a contrasting locality once and most schools, where Year 2 are separately grouped, will do so in this year. Where there are mixed age classes, of course, there is no option and the contrasting locality must be covered once in a two-year cycle.

In key stage 2, the minimum entitlement to the two contrasting localities is that they should be covered once, whether the school works annual cycles for the same year groups, two-year cycles for upper or lower juniors, or a series of whole-school geography topics differientiated according to age in multi-age classes. Many schools will, for the sake of a broad geography curriculum, want to provide access to two different contrasting UK localities and two different economically developing localities during the key stage. Where there are mixed age classes and problems with cross-mixing of classes from year to year, then localities in different countries can easily be used to prevent repetition. Schools which take their oldest pupils on a residential field trip will want to use this first-hand experience to develop contrasting UK locality learning after studying a different UK locality through second-hand resources lower down in the juniors.

Once places have been chosen and situated, investigate how you can mix in the themes with them. It is not advisable to treat the separate themes like parcels of geography. The presentation of the Revised Geography Order on the page suggests to some readers that a theme should be allocated to a Year, i.e.

- Key stage 1: 1 theme
- Key stage 2: 4 themes = 4 junior years.

This is not desirable because it does not allow for any revisiting of concepts relating to those themes at an older age. If you deal with rivers in Year 3, for example, even though you could use a progression of scales, for example local, national and European, it could make for a limiting geography curriculum where the nature of the learning, especially if you have no local stream to investigate, is by nature very tied to second-hand resources which at this age could be problematic.

When planning, consider combinations of places with a revisiting of some of themes as shown in the following examples.

Key stage 1
Year 1
Local area work
Investigate after children have an awareness of some local features: PoS 5a
What they find attractive and unattractive about the local park: PoS 6a
Fieldwork will be essential here: PoS 6c.

Year 2
Investigate where the school's waste is stored in the playground: PoS 5a
To where is it taken away? PoS 1c
What may happen to the rubbish tip when it is full? PoS 6c
Pupils will need to develop their geographical vocabulary: PoS 3a

Key stage 2
Here it is an example linking place and settlement:

Year 4
Local area geography study unit linked to local history unit of study
Developing the concept of the type of settlement you live in: PoS 9a
What it is like from the point of view of land use: PoS 9b, 5a
How its position gave rise to its growth leading into the historical aspects. You will be likely to involve:
Fieldwork: PoS 3b
Using maps and plans: PoS 3d
Secondary sources, e.g. aerial photographs: PoS 3e

Year 5
Geographical study of land use/potential environmental conflict issue: PoS 5a, 5d
Local area features and the issue of change in use of one or some of them: PoS 9c
The environmental consequences: PoS 10a, 10b
You will be likely to involve:
Fieldwork skill: PoS 3 b
Making plans and using appropriate keys and grid references: PoS 3c, 3d
Using newspaper resources: PoS 3e

Year 6
UK context, settlements: PoS 6
Different types of settlements in the UK, e.g. conurbations, coastal resorts, inland tourist towns building up atlas skills and OS smaller scale maps: PoS 9a
Developing map A awareness: PoS 3d
Appropriate terminology to name settlements: PoS 3a

In the examples above, notice that the PoS element is not just shown by its number and paragraph, or by the bold words of the programme of study. Your key stage plan will be most helpful if it includes the PoS numbers and paragraphs with the PoS words interpreted to a context which means something for your school. Otherwise developing continuity and progression will still be a problem, especially in key stage 2. For example, KS2 PoS 7b 'how rivers erode, transport and deposit' for both Year 4 and Year 6 could be taught with repetition if more specific context is not indicated, for instance the local stream and playground simulations in Year 4; a national river with aerial photographs and OS 1:50,000 maps, spotting meanders in Year 6. The key stage plan examples in Figures 4.2, 4.3 and 4.4 indicate the PoS and further content.

The extent to which you can develop theme work effectively will depend on:

- The time you wish to devote to the unit
- The nature of the geography of the place which you choose
- The nature of the resources which you have to address the teaching of that place.

If your economically developing locality (KS2) has no river running through or near it, then you will be unable to develop this theme. However, it would be inappropriate not to develop some aspect of the weather theme in this locality. This could be all or any aspect of KS2, PoS 8a, b, c depending on resources available.

In theory, you can consider likely skills PoS elements last of all, but in practice because as you think through and plan a key stage you will be bearing likely resources in mind, it is appropriate to add skills as you add the themes to the places. Skills need to be highlighted more precisely in unit planning and it may be useful to refer to Chapter 8, Figure 8.8 when considering the pitch of map skills in particular.

It is also worth noting that, when planning both a key stage and units of work, it is advisable to read through the level descriptions, not for assessment purposes but to have a feel for the pitch of work which needs to be planned. Remember that the average child will be:

- Working within level 2 at the end of key stage 1
- Working within level 3 by the time they reach transfer from lower to upper juniors
- Working within and likely to achieve level 4 by the end of key stage 2.

In summary

1. Decide on the nature of your key stage plan:
   - the type of cycles you have according to the nature of your school's age/class split
   - a range of blocked, continuous and linked units across half terms, terms, etc.
2. Cluster PoS elements of place, themes and skills together.
3. Use the PoS elements at least once and usually many times, especially skills.
4. Ensure that the PoS elements are given a context personalised for your school.
5. Check back to see that you are achieving reasonable progression in learning, through places, themes and skills.
6. Only plan what will fit into the time available.
7. Ensure you plan with the level and pitch in mind.

By mixing and matching the place elements of the PoS and the theme elements, it is possible to develop a very creative key stage plan, very personalised to your school and its needs. Now you need to agree the key stage plans with the staff if you have been asked to develop them. If they have been done with collaboration, then you will still need to check them over for continuity and progression.

The next stage is to expand each year's or cycle's work into a series of units of work. Class teachers may have to do this in your school, in which case it is your job as a coordinator to support and advise. In other schools, it is the responsibility of the subject coordinator to develop these units and discuss them with staff. This is an especially valuable approach where the subject coordinator is a genuine subject specialist.

Figure 4.2

Key stage 1 plan for geography aspects of topics: 36 hours work per year — **Context: An urban school**

	Reception		Year 1		Year 2
Autumn term	See Chapter 7: Early years geography Geographical language → Mapping through stories → Pre-plan view work → Working with globes → Routes: using, making, following → Enquiry →	Continuous reference to UK and world maps across the whole curriculum	*School locality, PoS 4* 6 hrs PoS 5a Physical and human features of school grounds – slopes, fences, etc. PoS 6a Sensory walk around school and grounds likes/dislikes of school environment PoS 3b Fieldwork PoS 3c Routes, trails, pathways, direction around classroom and school PoS 3d Plan of routes around school *Wider world context, PoS 1c* 4 hrs PoS 3e Globe and mapwork linked to Christmas customs in various countries PoS 3e Identification of land and sea PoS 3f Use of books and photos to explore images of these customs and their people and places	Continuous reference to UK and world maps as topical	*Local area, PoS 4* 10 hrs PoS 5a, b Land use and buildings around school: types and varieties of buildings; homes: detached, semi-detached, flats, etc; businesses; communal places PoS 3d Make maps of local area, key and plan view where possible PoS 3e } Use OS map and oblique aerial PoS 3f } photos to find features PoS 3b Fieldwork PoS 3a Use geographical terms PoS 3d Locate pupils' homes PoS 3c Make route maps from home to school *Local area, PoS 4* 4hrs PoS 6c Use *Dinosaurs and All That Rubbish* as a stimulus to investigate what happens to waste paper, glass and food from school site – interview caretaker
Spring term			*School locality, PoS 4* 8 hrs PoS 5c Effects of weather on school life – different uniforms and clothes, routines and activities according to weather PoS 3f Use photos to look at the effects of weather in PoS 1c different seasons around school and elsewhere *School locality, PoS 4* 6 hrs PoS 5d Land use of the school site – office, car park, caretaker's room, car parking, rubbish storage PoS 3d Contribute to class map of school site PoS 3b Fieldwork		*Wider world context, PoS 1c* 8 hrs PoS 5a Main physical and human features of a range of places and their wider contexts PoS 5c How weather affects people in them PoS 5d How land is used in them PoS 3f Use 'Watch' videos on Finland, etc., Oxford CD-ROM Infant Atlas, photographs PoS 5b Make some comparisons with own locality. *Wider context and local area, PoS 4* 4hrs In preparation for summer term, visit connected with science/geography linked visit: PoS 1c How local area is linked to other places children know or have heard about – other facilities they PoS 5a and their families use for work, shopping, leisure, visiting friends and relations. PoS 3e Mark where they live on local and UK map, name countries and features of UK
Summer term			*Local area, PoS 4* 6 hrs PoS 3b, 5a Fieldwork visit to local recreation ground to examine its physical and PoS 3b human features PoS 3c follow directions; introduce compass points; PoS 6a, b, c consider likes and dislikes and consider how to improve it *Linked unit with Science, PoS 1c* 6 hrs PoS 3c Fieldwork in a coastal place PoS 3a and 5a Recognition of physical and human features PoS 6a Views about the new environment PoS 5b Contrasts with own locality (3 hrs = fieldwork time from geography)		*Blocked – contrasting locality overseas 4* 10 hrs PoS 5 a, b, c, d, 1d Investigate through a second-hand resource an overseas locality. The amount of emphasis given to a particular PoS aspect will depend on the particular resource used. Comparison with local area is key. PoS 3d Use globes and maps to locate place.

Notes:
- Programme of study elements fit into wider topics unless blocked or linked is noted.
- Emphasis given to PoS elements will depend on time designated.

Figure 4.3

Key stage 2 plan for geography – minimum entitlement model: 45 hours work per year **Context: An urban school with many special needs pupils**

	Year 3	Year 4	Year 5	Year 6
Autumn term	*Local area* (largely to revise and consolidate mapping skills in new separate junior school) 20 hrs PoS 3b Fieldwork PoS 3c, PoS 3d } Maps and plans – make and use them on the school site and around the school (immediate area) PoS 3e Use aerial photos of area PoS 5a Identify physical and human features locally PoS 5d, PoS 10a } How the local area has recently been improved by the council PoS 2a/b Environmental survey sheets PoS 2c Communicate findings PoS 3d Introduce co-ordinates	*Settlement in Chatham* 20 hrs The whole urban area is considered here PoS 9a, PoS 9b } Settlement size, land use occupations, zoning PoS 3a appropriate vocabulary development PoS 5a (as 9a) Physical and human features of the town PoS 5c How the river, hills, change in dockyard use recently have influenced the town PoS 1a, 5e, 3d } How the town is linked to nearby settlements, other places in Kent and beyond – EU, UK PoS 3e Visitors to school, photos, brochures	*Weather in the UK* 10 hrs Continuous unit over 3 months of 2 terms investigation PoS 3b Use weather instruments to record local weather PoS 3c Take a range of weather PoS 8c data for London, Glasgow PoS 6 Edinburgh, Cardiff, etc. from secondary resources PoS 3f Use data handling IT to log PoS 2 measurements then pull all work a/b/c together to analyse data collected and display it PoS 3f Use weather forecasts, photos to PoS 8b understand seasonal change PoS 3c Locate places on atlas	*Contrasting developing locality* 20hrs PoS 4 Kenyan village and Kenyan city PoS 5a Main physical and human features PoS 5b Similarities and differences between Nairobi, Kaptalamwa and own locality PoS 5c Features of Kenyan places PoS 5d, PoS 1d } Locating Kenya, its localities and links with wider area PoS 9a, PoS 9b } Use Nairobi and Kaptalamwa to illustrate more about settlement theme PoS 8b Weather in these places PoS 3e Use secondary resources PoS 3d Using maps and globes
Spring term	*Our local river* 14 hrs including visit PoS 7a, PoS 3b } River system features based on real experience down in town centre to view river and river beaches PoS 3e Use photographs and rivers video PoS 7 b to clarify observations and introduce erosion and deposition PoS 10a How people pollute rivers – through on-site observation	5 hrs *Local area history study unit – Chatham in the past* This unit could be swapped with autumn term geography unit according to the number of classes/resources school possesses or most logical order of concepts for pupils before summer term geography unit PoS 3d Mapwork in the context of history	*Settlement in the UK and EU* 15 hrs PoS 9a, PoS 3a, PoS 3e } Villages, towns, cities, etc. appropriate vocabulary use 'Settlement' videos, range of aerial photos to understand types of settlements and to be able to identify them PoS 3e Be able to locate settlements, e.g. PoS 6 through well-known football teams or holiday resorts in UK, EU using atlas, road atlas and EU country map examples. Draw maps to design a visit to UK or EU area	*Rivers* (unit linked to history study unit on Egyptians 15 hrs PoS 7a The River Medway system and the River Thames PoS 3a Use and interpret OS 1:50,000 maps and atlas maps, contents page, index, revise coordinates, 4-figure grid references, aerial photos of rivers PoS 7a The *modern* River Nile, its features, PoS 3d e.g. delta, and use to *modern* Egyptians using appropriate atlas maps and vocabulary
Summer term	*Micro climate in the school grounds* (Continuous unit) 6 hrs PoS 8a Build on autumn term work by investigating and mapping microclimate in the school grounds (use base maps from autumn) PoS 3b Use instruments to measure wind direction, temperature, etc. PoS 2 a/b/c Collect data, analyse data PoS 3f Collect data on IT database	*Contrasting UK locality* 20 hrs PoS 3b Fieldwork visit to village 5 miles away PoS 9a, PoS 9b } Settlement size, location Economic activity, land use in Burham PoS 5a Its main physical, human and PoS 7a environmental features, including River Medway PoS 5b Similarities/differences with Chatham PoS 5c How its features (river location, etc.) influence its businesses PoS 5e, PoS 1d } How Burham is linked to the rest of Kent and UK PoS 2 a,b,c Collect, record, analyse data	*Local issue – The Medway tunnel* 20 hrs PoS 9c, PoS 10a } The construction of tunnel under the River Medway and how it will PoS 5a change the environment PoS 3d Examining maps and publicity PoS 3e about the project seeking PoS 3b peoples' opinions through questionnaires	*Global environmental issues* 10 hrs PoS 10b Research into a range of ways in which we try to manage environments, e.g. river pollution, rainforests, topical examples where possible PoS 3e Secondary enquiry using videos, PoS 2a/b/c library books, topical articles at a range of scales PoS 1a PoS 6

Notes
- Units are blocked unless otherwise indicated.
- Work is grouped within terms – it may occupy a whole term, half a term or less according to number of hours.
- PoS elements will be covered in varying depth according to time available.

Figure 4.4

Key stage 2 plan for geography – maximum entitlement model: 45 hours work per year Context: A suburban school with adjacent countryside within easy access

	Year 3	Year 4	Year 5	Year 6
Autumn term	*Local Area* 15 hrs PoS 3b Fieldwork visit to local farm to investigate land use of the area and what the farm does Pos 1d, PoS 5e } with its produce – where it goes, how it travels there PoS 10a Investigate whether farming affects the area and how clearing hedges, the need to manage PoS 2a, b, c pests, heavy lorries transporting goods affects it PoS 5a The physical and human features of the local area	*Wider local area* 15 hrs PoS 3b, PoS 3d, PoS 3e } Through fieldwork, using aerial photos and maps and by making maps PoS 9a, PoS 5a } Investigate the local settlement, its quarry, shops and city centre area PoS 9a, PoS 3e } Use 'Settlement' video to build up the notion of different types of settlements and the right vocabulary PoS 6, PoS 3e } Use atlases and UK map to find and plot different examples of settlement types children have heard of or know	*Contrasting urban locality in developing country: Cairo and 4 localities within it* 15 hrs PoS 5a, PoS 5c } The physical and human features of different parts of Cairo, its site and situation PoS 5c Comparing Cairo and own area PoS 9a, PoS 9b } Different parts of Cairo – land use: residential, motorways, markets, etc. PoS 10a, PoS 10b } How residents look after their areas, how the government looks after it PoS 3e, PoS 3f Photopack and video of Cairo produced by Oxfam, reference books, Encarta PoS 3d Using maps and globes PoS 7a The Nile River System	*Rainforest environment* 15 hrs linked with a Science Unit of work PoS 8c Weather conditions in different parts of the world – a case study: the Amazon PoS 10a How people locally, nationally and globally affect the rainforest PoS 10b and thus how its maintenance and sustenance affects us PoS 3e Use of Landmarks Rainforest video, reference books on Brazil, Amazonia, CD-ROMs PoS 3f World Atlas, Distant Places Nelson Primary Atlas, Encarta PoS 7a, PoS 7b } Use photos and atlases to investigate the River Amazon
Spring term	*Streams and rivers* 15 hrs PoS 3a, PoS 7b } Use of school grounds, sand tray and guttering simulations to illustrate the concept of erosion and deposition and some features of a river system PoS 3e Use of 'Rivers' video to illustrate river features Use of photopacks of river features PoS 3a Development of geographical terms PoS 7b, PoS 3c } Field visit (towards end of term) to local stream, take photos and measurements	*Contrasting UK locality: an urban locality through a link with a twin school and secondary resources* 15 hrs PoS 5a Main physical and human features of locality PoS 5b Comparing it with own locality PoS 5c How the features of the locality affect life there PoS 5e, PoS 1d, PoS 6 } How to travel to and from own locality, its links with National Park and rest of UK PoS 3e Using snapshots of features from twin school and its area. PoS 3f Compiling questionnaires to send to twin school PoS 3d, c Making and using maps, plans	*Weather locally, and in UK* 15 hrs PoS 8a Ongoing throughout the term – micro climate measurements in the Pos 3c school grounds using fieldwork instruments PoS 8c, PoS 8b, PoS 3e, PoS 3f Weather conditions in the UK – as appropriate use weather forecasts, videos, teletext details, newspaper cuttings to build up a picture of winter UK weather. Contrast with extremes of weather as/when they occur elsewhere in Europe or world PoS 6 Use maps to identify places mentioned. Map school grounds to plot sites for micro-climate work	*Contrasting UK locality: residential fieldwork visit* 30hrs Also developing history-linked unit Visit early summer term, thus unit allows for preparation and follow up over 1½ terms PoS 3b Fieldwork, questionnaires, landscape sketching, compass use
Summer term	*Contrasting rural locality in a developing country* 15 hrs Village in the Swat Valley, Pakistan PoS 1d, PoS 5e, PoS 3d } Getting to Pakistan – using the world map and globe and atlas increasing map C awareness; locating the locality PoS 3e, PoS 5a, PoS 5b Use Swat Valley photopack to develop knowledge of physical and human features of the village in its area PoS 7a, PoS 7b, PoS 8a, b, c Develop some idea of the importance of the river and weather in the area	*Investigating the EU* 15 hrs PoS 1d, PoS 5e Using children's own experience of holidays or news items or products from EU – build up awareness of map of Europe, routes to continental Europe, Channel Tunnel, Eurostar, etc. PoS 6 Investigate, with a geographical framework one or several countries of Europe as individual group or class projects. Plot European rivers Pos 3e Use relevant video sections, library books, maps, atlases and PoS 3d globe, contents, index of atlas	*Local land use issue* 15 hrs PoS 2 a/b/c Geographical enquiry into the proposal to build an adjacent supermarket PoS 5d Proposed change in the local area PoS 9c How people feel about the proposed plan – the old, mums, children PoS 3c Fieldwork questionnaires PoS 3c Make and use maps of the Pos 3d proposed site and local area	PoS 5d How fieldwork locality relates to its surroundings PoS 1d Isle of Wight linked to school area PoS 6 UK context of Isle of Wight PoS 5a Main physical and human features of fieldwork locality, i.e. around the hotel or hostel, beach work, housing survey PoS 10a Linked to the environmental PoS 9c issues of tourism in the resort PoS 5b Comparisons with school area PoS 3d Using OS maps on fieldwork using a range of maps in preparation and follow-up

Notes
- Units are blocked unless otherwise indicated.
- Work is grouped within *terms* – this means it may occupy a whole term, half a term or less according to number of hours.
- Not all PoS elements can be covered in depth – the time for the unit is the key.

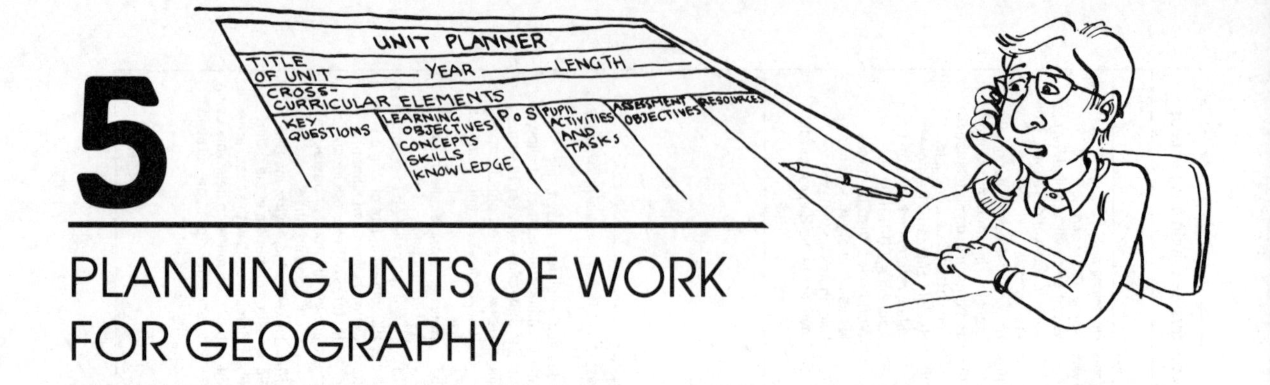

5

PLANNING UNITS OF WORK FOR GEOGRAPHY

Introduction

Every school should teach geography within a framework which ensures clear progression. Whether you call the geographical element of work a topic, project, study unit or integrated studies, it has to be planned. For the purposes of this chapter, we will use the term 'unit of work'. The length of a unit is best defined for official requirements in terms of hours taken, but this must not become a constraint. The same results can be achieved by a class working for one morning and three afternoons a week for two weeks, as by a class working one afternoon a week for eight weeks. A flexible approach to which subjects should, or could, be taught with geography should be considered when planning units (see Chapter 3).

The medium-term planning of units of work is the next step. You will have been given various PoS to be covered in a term, half-term or year, and it is up to you to decide how best to turn those PoS into units of work which suit your class and your own personal teaching style. Look at the whole primary curriculum for the term or half-term and decide how much integration or linking with other subjects is desirable or possible. Remember to think about the cross curricular themes of environmental education, health education, economic and industrial understanding, citizenship and careers education as well (see Chapter 12). The amount of integration or linking will vary over the year depending on the structure of the science, history and design technology key stage plans. To a lesser or greater degree the links with maths and English will always be there. It is worth stressing that variety in the duration and depth of units of work and the integration or linking of geography with different subjects at different times should lead to a broad, balanced primary curriculum. You may wish to teach:

- Termly units
- Half term units
- Short focused units of a few weeks
- A broad unit of work incorporating many subjects
- A linked geography and science unit
- A linked history and geography unit with a continuous science unit running parallel
- Three separate units running parallel or in succession
- Any other combination that fits with your circumstances.

Your decision must be guided by school policy and the fact that different teachers work best in different ways. The nature of your class and the school's long-term plans will also lead to work being developed in different ways. When planning, remember that there is of course the issue of equal opportunities across parallel classes which needs to be considered. The unit-planning approach described here can be used for individual subjects or for integrated planning. It provides the head, coordinator or school with a copy of the medium-term planning related to the National Curriculum and assessment. It also provides the teacher with a working document to use throughout the unit of work as a lesson planner and, at the end of the unit, it can be used to evaluate – thus saving everybody's time. This unit planning approach can be used equally well for history, science or technology. It is often easier for teachers if the same planning format is used for all these subjects within a school.

Beware if you have just been given a list of topic or unit titles to cover over the year. It is possible to teach familiar titles such as 'My Home', 'Houses', 'Underground', and so on, without teaching the PoS required by National Curriculum geography. It is very comfortable to see favourite topic names and you could be excused for feeling that it is alright to carry on teaching them in the way you always have. But, if you do it is likely you will teach without covering effectively any of the PoS. So, when you want to make judgements about children's progress in geography, or fix on particular aspects to assess, you will be unable to do so effectively. A lot of work will have been done by both pupil and teacher, but it will not be sufficiently relevant to the National Curriculum or geography itself. It is only through working from the PoS and identifying clear learning objectives that National Curriculum geography can be delivered. Titles can be reused, but the content of the unit must be directly linked to the PoS and, through them, to formative and summative assessment.

Planning the achievable

Time is the most precious commodity in the primary classroom; by planning only that which you can achieve, you save time. The 'topic web' approach to planning, which was based on brain storming around a title or theme, generated lots of work and activity in many areas of the curriculum which was integrated to a greater or lesser degree. You then taught as much of this as you were capable of in the time available. This often led to a sense of frustration 'I never got round to ...', or 'I wish I had time to do ...', or 'How can I fit geography in to this?' The web-planning approach often involved the preparation of a list of content or general approaches with no reference to concepts, skills and knowledge. Topics were often unfocused and the potential was there for them to become more complex as more National Curriculum subjects came on line by 1993. Within the Revised National Curriculum we have to be much clearer in our thinking as to what geographical knowledge, understanding and skills we need to teach and when (see Figure 5.1).

Figure 5.1

Sequence for planning units of work	
PoS	Read the PoS required to be taught, and note them in numbers and words.
Questions	Formulate focus questions targeted to PoS.
Learning objectives	Devise the learning outcomes you want related to the PoS and key questions.
Tasks	Devise pupil tasks to encourage and produce evidence of pupil understanding.
Resources	Make a note of the resources you have and any you need to collect.
Assessment points	Highlight any assessments to be recorded and acted upon.
Links	Add any cross-curricular links and any link to the cross-curricular themes, and/or dimensions.

Your starting point is the PoS you have been asked to deliver. Read them through and see how you think they will best fit together to make a coherent unit of work. Then try to focus the main ideas from the PoS into questions. By using the enquiry approach you will raise questions which will lead the learning process. The children's completed tasks will produce evidence to show you whether they understand the concepts and knowledge and have mastered the skills necessary. This means that the children's completed units of work can be used for assessment purposes as evidence of their progress. Assessment does not have to be tacked on to the end of every unit (see Chapter 6). Figure 5.2 gives some examples of key questions for various key stage 1 and 2 units of work. These are by no means the only questions you could use for these units. Remember to plan only what you can achieve in the time you have: you may only be able to cover four to six questions in half a term or six to ten questions in a full term. Make allowances for a short calendar term or for Christmas taking time from the term. With experience you will get to know the amount of work and time needed: it could take a couple of weeks' work to answer one question, whereas another could be answered in half an hour.

Think about the tasks you will ask the pupils to do:

- Will they be working in groups, alone or as a class?

Figure 5.2

Key questions for grid planning

People who help us KS1

PoS 4 Local area
PoS 5a Features of the local area
PoS 5d Land and building use

What are the different parts of our school?
What are they used for?
Who works in them?
What buildings do we know around our school?

Seaside KS1

PoS 4 Contrasting UK locality
PoS 5a Main physical and human features
PoS 5b Similarities and differences with local area
PoS 6a Environmental aspects of the beach

How is the seaside different from your own local area?
What is the land shaped like at the seaside?
Do you know any special words to describe the land at the seaside?
What buildings do you see at the seaside?
Can you design a coastal landscape?
Do people damage the seaside?

Our area KS2

PoS 4 Local area
PoS 9a Settlement size, character and location
PoS 9b Land use
PoS 5a Human and physical features
PoS 3c Make maps and plans
PoS 3d Use maps and plans
PoS 3e Use secondary sources

What is the land use for in our area?
How can we map the land use?
Is it mainly residential, commercial, recreational or industrial?
Can you see a pattern in the land use?
When did the settlement in our area develop?
What type of settlement is it?
Is there a pattern to the settlement?

Chembakolli (India) KS2

PoS 1d Wider geographical context
PoS 5e Broader context and links with elsewhere
PoS 4 Developing locality
PoS 5a Main features, physical and human
PoS 5b Similarities and differences
PoS 9b Land use, e.g. farming
PoS 7b Weather – seasons
PoS 7c Weather conditions elsewhere
PoS 3e Use secondary sources – photos
PoS 3d Use atlases

Where is India and how big is it?
Do all parts of India have the same weather?
What the main landscape features of India?
Where is Chembakolli?
What do the photos tell us about life in Chembakolli?
What happens during one day in Chembakolli village?
What might happen during one day in our local area?
What is the weather like in Chembakolli?
How does the weather affect peoples' life in Chembakolli?
How does Chembakolli compare with our local area?
What ten things would you send to Chembakolli to tell them about our area?

River Thames KS2

PoS 7a River features and systems
PoS 10a How people manage the environment
PoS 8b Seasonal weather patterns
PoS 3e Secondary sources
Pos 6 UK context

Where does the River Thames start and finish?
What are the Thames' vital statistics?
As you travel along the Thames from W to E what changes would you see?
How can the River Thames affect people's lives?
Do people have an effect on the river Thames?
How and where can you get to the other side?

Note
Welsh Order key questions need to be different as PoS are slightly different.

- What do you expect the outcome to be?
- How will you teach?
- What resources do you need/do you have?
- How will you differentiate – by task, resources, teacher intervention or outcome?
- How will you deal with children who have special needs?

Remember, as you decide what tasks the children will do, to build in some formative assessment points (see Chapter 6). If the tasks are targeted towards the programme of study and through them towards the specific learning outcomes, then evidence of achievement will be available for you to use for teacher assessment and for you to record progress and act on.

Fill in the resources column on the unit plan, listing both what you know is available and what you will need. Some of the activities you plan should tell you about children's progress informally, perhaps one in a unit should enable you to make a summative judgement related to, say, the child's understanding of river processes. Highlight this task in the assessment column with an asterisk. Cross reference the unit by putting the PoS and pitch of the work related to the appropriate level in the correct columns.

Note any cross-curricular links in the appropriate column, unless you are designing a linked unit, when they will be shown. Also make reference here to any links with the cross-curricular themes of environmental education, health education, economic and industrial understanding, citizenship, careers and work education. Links to other dimensions considered by the school should be highlighted, for example European dimension, multi-cultural education.

Unit planning grids

Two examples of blank planning grids are shown in Figures 5.3 and 5.4. These grid plans will help you to focus your planning and develop classroom tasks that are targeted towards the National Curriculum. There are many forms of these grids, but the annotated one (Figure 5.4) is the one the

Figure 5.3

Unit planner					
Title of unit............................	**Year**		**Term**.................		**Length**....................
Cross curricular elements					
Key questions	**Learning objectives** Concepts Skills Knowledge	**PoS**	**Pupil activities**	**Assessment objectives and tasks**	**Resources**

authors have found most useful for unit planning in key stages 1 and 2.

Some examples of completed grids are shown in Figures 5.5, 5.6, 5.7 and 5.8. The completed grids show examples of planned units of work for both key stages. They are not the only way of tackling these units, just one way, using an enquiry-based approach.

Differentiation

Differentiation is central to the process of learning and assessment. Ideally, it should be built into the units when you are planning the children's tasks. You may find you are only able to do this in medium-term planning after much experience if you are not a geography specialist. If you cannot manage it in medium-term planning, ensure you do it in short-term or lesson planning. It may be by task or outcome. All children learn at different rates, in different ways and to different levels of achievement. The fact that they need to be given equal opportunities to achieve must, however, be kept in mind when planning their tasks. We need to offer the children tasks where they can demonstrate their progress. When planning units of work the relationship between the task and the focus question is important because that targets the children's work directly to the National Curriculum. The levels and types of differentiation are equally important for positive pupil achievement and equal access to learning experiences.

Differentiation by task is where pupils working on the same section of a unit of work are given different tasks. These tasks may require different levels of support and resources. This type of differentiation can also involve a series of structured or stepped tasks; these are usually open-ended and children can progress through them.

Differentiation by outcome is where pupils are involved in the same task; the work they produce indicates their level of achievement.

This example shows differentiation by task and outcome in a local area study, looking at the form and features of the settlement. The mixed-age-group class of seven- to nine-year-olds was going to a local viewpoint, a hill in their rural locality. The class worked in three ability groups.

Task

Group A

Pupils are given a prepared sheet showing two boxes, one in which to draw their home and one in which to draw their school. They have to join the boxes showing their route to school and put in three landmarks along the way.

On a second sheet two boxes showing the school and the viewpoint must be connected by drawing the route and three landmarks whilst in the field.

Group B

Pupils predict the route and check it in the field. They mark, on a photocopied O.S. 1:2500 map, the route they expect to take to get to the viewpoint, and indicate certain features the teacher asks for. En route to the viewpoint they check and amend their maps.

Group C

As for group B, then colour code the buildings along the way according to use and type, for example, farm, shop, detached house.

Outcome All groups were asked to do a landscape sketch from the viewpoint, naming as many features as they could. This work was then differentiated by outcome as shown in the pupils' work. It could also be differentiated by the resources offered to the pupils or by teacher intervention through constructive questioning.

Figure 5.4

Unit planner					Type	
Title		Time .... hours		Term	Class	
PoS	Key questions	Learning outcomes	Tasks	Resources	Assessment points	Cross-curricular links

The title of the unit

The code and words for the PoS included in this unit

The main questions upon which the pupils will base their work, developed from these:
Where is this place?
What is this place like?
Why is this place like this?
How is this place changing?
How is this place connected to other places?
How is this place similar to or different from another place?
What is it like to be in this place?

The intended geographical knowledge, understanding and skills to be taught and learned in this unit

The activities which will be covered in the lesson, describing the learning process and teaching methods. Activities should be planned to allow pupils to work at their own level. The outcomes of the tasks can form the evidence for teacher assessment.

The resources you would need for each individual lesson or section of the work. Don't forget IT.

Don't panic about assessment. Not all work covered in the unit needs to be assessed. Your coordinator will have told you of any isolated points that must be assessed at this stage.

Links to other areas of the curriculum and any links to the five cross-curricular themes of environmental education, economic and industrial understanding, health, citizenship and careers. Dimensions such as multi-cultural education may also be included.

Indicate whether this is a blocked, linked or continuous unit of work

Figure 5.5 Example of a linked unit of work

Geography and science: Local area or contrasting locality in UK					Linked
		Year 1/2	Time: 12 hours geography	Term: Summer	
PoS	**Key questions**	**Learning outcomes**	**Activities**	**Resources**	**Criteria for assessment**
1a Investigate physical and human features of surroundings 1b Studies which focus on geographical questions	What do we know about the seaside?	To begin to share knowledge and ask focused questions	Pupils suggest what they already know, listed by teacher Pupils suggest a question they would like to find out the answer to Preparation for day visit to the seaside, local or distant, highlighting safety, grouping, behaviour, organisational items	Large stimulus photo – big book or Collins–Longman A3 photo	
3a Use geographical terms 3b Undertake fieldwork	Who works at the seaside? Who plays at the seaside?	Focused observation of activities and features	Visit to harbour area or equivalent. Observation of any work evidence – fisher boats, ferries. Observation of leisure activities – pleasure boats, windsurfers, etc. Count how many of each type	Local or distant seaside location	
4 Local area or contrasting UK locality visit 5a Main physical and human features that give the locality its character			Visit to beach area – further observation of leisure activities, beach features, cliffs, environmental issues (flotsam, jetsam cast up, rubbish dropped, etc.)	Beach and coastline – variable with cliffs, pebbles and according to area	
6a To express views on attractive/ unattractive areas of environment 5d To know how land and buildings are used	What do we like about the seaside?	Sensory experience of basic natural features – sand, rock pools, pebbles, etc.	Supervised free play on the beach. Pupils collect 3 or 4 safe objects to take back to school in a plastic bag Listen to sounds for a minute Children articulate their likes and dislikes Recall work being done according to ability/development, list these by drawing and/or writing. Recall leisure activities in the same way Graph the numbers of boats, yachts, ferries seen	Natural environment Photos and commercial postcards (if available) Snapshots of the features and activities taken on site are essential	Children engaged with the real environ-ment, enjoying themselves *Assessment point Being able in recall/record some features and activities done at the seaside

Adapted from work originally done by the author for Kent County Council

Figure 5.6 Example of a continuing unit of work

Geography around the UK					
		Year 3	Time: 5 hours	Term: On-going	
PoS	**Key questions**	**Learning outcomes**	**Activities**	**Resources**	**Criteria for assessment**
1a Investigate places across a widening range of scales	Examples: Where do my relations live and how do I get there?	Increasing awareness of knowledge of places, features in the UK and links between them: motorway links, trader links, family links	A series of small enquiries based on a mixture of topical and appropriate issues within the class and/or arising through the news, over three or four times a year either when there is little other geography in a term or half term, or when it fits well with a blocked unit of geography	Wall maps of UK, atlas maps of UK and parts of UK. Any other maps which pupils can relate to	Ability to find places on the map and suggest a route
1b, 1c, 1d Become aware of how places fit into a wider geographical context	Where have we been on holiday and how did we get there?				Sketch map sequencing places or making links. Filling in map provided. Basic comparisons made
2a, b, c Low level enquiry	Where did my family live before? What was it like there?				
3d Use and interpret maps and plans at a variety of scales	What is the weather like in Scotland compared to here?			TV weather forecast or reports, newspaper reports	
Possibly 10a How people affect the environment	Where are the polluted beaches?			Information from local site visits for geography or other curriculum areas	Appropriate routes indicated or drawn * Assessment point – increased place knowledge
	Where do the strawberries from the fruit farm go to?		Using information obtained locally to plot outline routes		

Adapted from work originally done by the author for Kent County Council

Figure 5.7 Example of a blocked unit of work

Week		Spring term geography: Blocked unit of work, Years 5 and 6 (2 parallel mixed-age classes, 6T and 6C) 16 hours					
6T	6C	Key questions	Programme of study	Learning outcomes	Pupil activities	Resources	Assessment points
2	3	What is a river?	7a River features and characteristics 7b Erosion and deposition	Describe the journey of a river – the patterns it makes and processes which form it	• See video • Sketch journey of a river with labels • Make large class collage for display	Gogglebox 'Rivers' video UK Atlas, page 9	
3	4	What are the Rivers Medway and Thames like? How do they change?	7a River features and characteristics 3e Use secondary evidence sources	• To identify river features on maps and photos • To use aerial photos and maps	• Studying and sorting aerial photos (guided by teacher-made worksheet)	Aerial photopacks of Medway and Thames 1: 50,000, 1: 25,000 OS maps	
4	5	Which are the most famous rivers?	6 UK and EU context, national and global scale 3d Use and interpret globes and maps	• To identify and locate some rivers of Britain , Europe and the world	• Maps of the British Isles and world – for pupils to label rivers	Published prepared sheet maps of the British Isles, Europe Atlases	(level 4/3)
5	6	What is a dock?	7a River mouth or estuary features 3d Use and interpret maps	• To understand why rivers are important • To know about the use and change in use of rivers	• Map skills task on the Albert Dock, Liverpool • Discussion on London Docklands	'Master Maps' page 14 & 15 Aerial photos of River Thames, Satellite image of Docklands 'East Enders' title map	* Map skills assessment point
6	7	Should rivers be used for power?	7a River systems 2a Observe and ask questions about issues 2c Analyse, conclude	• To have some knowledge of the importance of rivers • To know about issues related to river system management	• 2 x ½ class size debates on issues – preparation and presentation of debates	Commercially published resources on Severn Bridge and Narmada Dam (India)	(level 4) English* speaking and listening
7	8	What are locks? Could the River Thames flood London?	7a River systems and their features 10b Managing environments	• To understand the dangers of rivers and how humans intervene	• Group work research and presentations to class	National Rivers Authority pack Sheet on locks Aerial photos Thames book	As above
8	9	*Fieldwork* What river features can you see at Eynsford and at Allington Locks? Joint field trip with history to Roman villa.	7a River systems 7b River features River patterns River processes 2a Observe and ask questions 2b Collect and record evidence 3b Fieldwork	• To identify river systems and compare the two sites	• Field sketches • Locating rivers on a range of maps	OS maps 1: 50,000 1: 25,000 1: 10,000 extract copied sheets	Assessment point * PoS 7a
12	12	*Fieldwork* Local stream study in Cuckoo Woods How fast is the water moving? Can you see evidence of erosion and deposition?	7a River features 7b Erosion and deposition Joint trip with science also to Court Lodge Farm and Boxley Church – Rocks and soils.	• Understand speed of flow, erosion and deposition	• Measure speed of flow • Photograph features	Dog biscuits, tape measures, cameras	Assessment point *PoS 7b

Source: Sandling C.P. School, Kent

Figure 5.8 Example of a blocked unit of work

Geography: St Lucia and locality within it, Castries					
Year 6		Time: 20 hours		Term: Spring	
PoS	**Key questions**	**Learning outcomes**	**Activities**	**Resources**	**Criteria for assessment**
4 Locality in the Caribbean	What do we know about St Lucia?	To raise awareness of the place	Brainstorming knowledge, lack of it, opinions.	Blackboard or paper	*Pupils' knowledge self assessment
1a, 1b, 1d 3d, 3e, 3f Secondary resources	Where is St Lucia and its settlements? and how do we get there?	To locate St Lucia and know how long it takes to fly there	Locate the Caribbean, St Lucia and its principal settlements and airport. Find out how far away it is and how long it takes to fly there.	Atlases, world wall map, satellite map of St Lucia, Tourist literature, possible CD-ROM encyclopedia use:	Knowledge of location of island
5a, 5b, 5c, 9b, 9a	What are St Lucia and Castries like? (Pupils come up with their own questions to contribute to this main one)	To have a general overview of what St Lucia is like	Discussion using photographic evidence and written notes which accompany them. Written work which systematically structures what the physical and human environment is like.	Distant places Photographs from GA St Lucia pack	Pupils able to come up with questions
5a, 5b, 5c, 8b Seasonal weather patterns. 8c Weather conditions in different part of the world.	Why is the island like it is? a) temperature b) rainfall	To understand that St Lucia is tropical because of its proximity to the equator. To appreciate the weather patterns in St.Lucia in comparison to the local area's weather patterns	Produce graphs of temperature in St Lucia and London Make and compare graphs of St Lucia with school's own weather records.	*Primary Geographer*, No. 9, P7 IT database	Oral and written comparative analysis and conclusion statements Pitch – level 4
Science AT3/2e Geog 10b	c) water cycle (science)	To reinforce the water cycles and understand that water is a precious resource.	Teacher explanation of water cycle using diagrams and/or practical demonstrations.	*Primary Geographer*, No. 9, P7 **Note:** Dealing with volcanic features on the island is an essential part of its place character, but this should not turn into a major aspect as volcanoes and earthquakes in a KS3 theme.	Pupils' diagrams and explanations
5a The main physical and human features and environmental issues.	d) volcanic features	To know what a volcano is and how it affects people	Identification of volcanic areas and scenery on the island using photos and maps. Look at and learn about cross section of a volcano. Explain how volcanoes help or hinder people.	Map of St Lucia, photo pack, atlas, information books, photocopied worksheets.	*Assessment point Here/there chart – completion of chart giving comparisons and explanations.
5b How the localities are similar and how they differ.	What is it like to live in Castries and St Lucia?	To understand how people live in Castries	Comparison between lifestyles on Castries and in own locality.	*Geography starts here* Video Pen Pals: St Lucia. ITV Spring 95. Range of photos from St Lucia pack. Knowledge and photos of school locality.	Compare initial, mid point and final knowledge to judge progress in knowledge and understanding of places
5e How the localities are set within a broader geographical context.	What is exported from Castries?	To understand that banana growing is important to the island economy	Use port photos; hypothesise. Examine various aspects of the banana industry on St Lucia – inputs, outputs, cultivation cycle, use of diagrams.	Photo and slide resources. *Primary Geographer*, No. 11	
	How does tourism affect the island?	To understand the importance of tourism to a developing country	Produce profiles of local people employed in, various aspects of the holiday trade (hotel owner, chambermaid, diving instructor, etc.)	Tourist brochures with images of St Lucian reports/environments. TV video holiday programme clips.	
10a How people affect the environment.		To understand the environmental threats of tourism	Role play – groups take the role of hotel developer, local banana grower, conservationist, tourist, etc.		
10b How and why people seek to manage and sustain their environment.		To make what efforts are made to manage the threats to the environment	Simulation of public enquiry into the building of a five star hotel complex with several swimming pools, large gardens, in an exceptional site.	Previous knowledge learned, ideas understood so far., access to any resources so far. *Primary Geographer*, No 22 St. Lucia special July 1995 is an additional resource for all this work.	Teacher listens to reasoning showing awareness of more than one issue *Assessment Pupils write a letter to a relative in their locality summarising their knowledge of Castries and St Lucia

Note: Dealing with volcanic features on the island is an essential part of its place character, but this should not turn into a major aspect as 'volcanoes and earthquakes' is a KS3 theme.

Adapted from work originally done by the author for Kent County Council

Conclusion

Remember to keep things simple when you are planning units of work. Only plan that which you think you can achieve in the time provided. Start from the PoS and integrate different aspects of them where possible. If there is any extra time, build in other good geography activities or use it for extension tasks and reinforcement.

Remember that the unit of work plan should also act as a class record – the coordinator should have a copy with any areas *not* taught highlighted for quick visual access. Evaluative comments at the end of, or on the back of, the unit as to its strengths and weaknesses will assist better re-use of the unit another time.

6

ASSESSMENT: NOT THE BOLT-ON EXTRA

Assessment in geography

Teacher assessment is not a new task; we have always assessed children informally to be able to plan the next steps in their learning. We observe pupils, ask open-ended questions, listen to children, look at their work, and use the knowledge gained to judge where the children are in their development and what the next step is for each pupil.

Assessment is an essential step in the teaching and learning cycle. Teachers need to be clear of the learning outcomes they expect at the different planning stage. They should be evident in different ways in their long-, medium- and short-term planning. Assessment should support the planning and feed back into it. There need to be strategies for those who progress at different rates, either faster or slower than the majority of the class. This differentiation has to be based on the every-day teacher assessments and fed back into the continuous process of planning, teaching, learning, assessment, planning and so on. This link is clearly shown in Figure 6.1. Good assessment practice supports good teaching and learning. This is not a new idea – in 1988 the Task Group for Assessment and Testing (TGAT) said:

> "*Assessment is at the heart of the process of promoting children's learning. It can provide a framework in which educational objectives may be set and pupils' progress charted and expressed. It can yield a basis for planning the next educational steps in response to children's needs. By facilitating dialogue between teachers, it can enhance professional skills and help the school as a whole to strengthen learning across the curriculum and throughout its age range.*"

Teachers need to assess in order to know that a pupil has learnt. The next step in the learning process cannot be planned if you have not informally assessed where the pupil is now, after the latest imput or activity. So assessment should not be viewed as some separate formal activity, but as an integral part of your teaching.

These are the general principles of assessment that a primary school should consider in planning its assessment policy:

- Assessment must be an integral part of teaching.
- Assessment must provide all pupils with the opportunity to demonstrate achievement.
- A variety of techniques should be employed so that the assessment is appropriate.

Figure 6.1

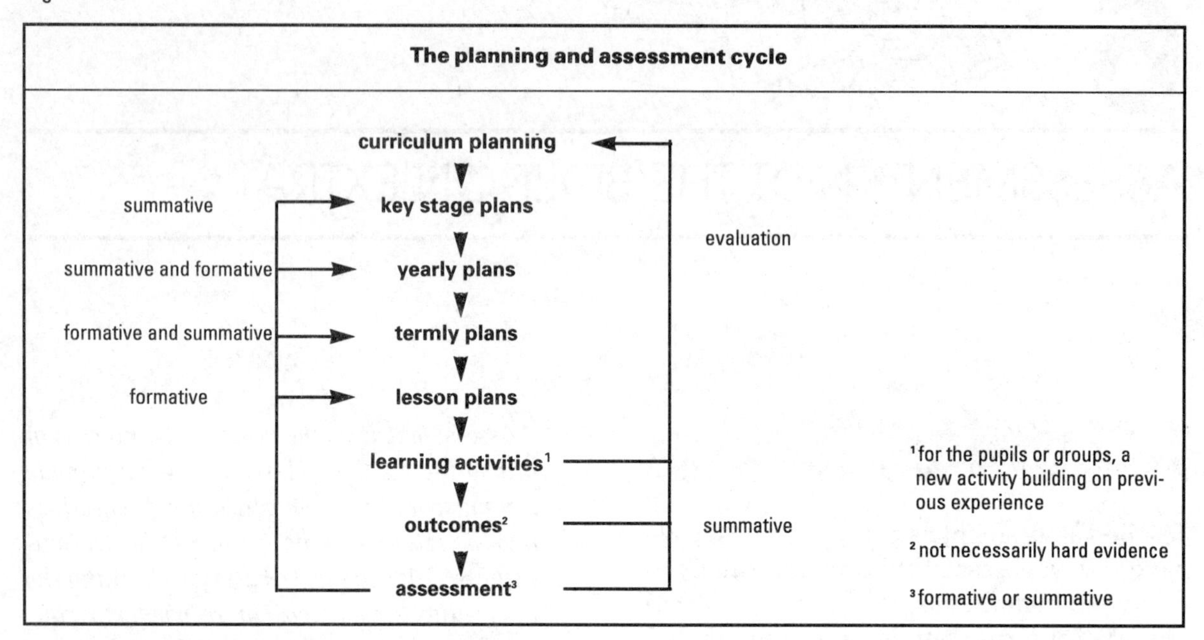

- Assessment strategies and standards should be agreed by all teachers involved in the assessment.
- The opportunity should be available to assess when ready and reassess if necessary.
- Assessment should involve the pupil.
- Assessment should be enjoyable.

All of the principles listed above apply to assessment in geography.

Assessment and the Revised Order for geography

Within the Revised Orders for the different National Curriculum subjects, the formal summative assessment arrangements have been streamlined and simplified. DFEE have also placed confidence in teachers' assessments and acknowledged the crucial role teachers play in assessing their pupils. In *Consistency in Teacher Assessments, Guidance for Schools Key Stage 1 to 3* (SCAA, 1995), there is a clear definition of the dual role of teacher assessment:

"The term 'teacher assessment' is commonly used to describe everyday assessments which take place through a key stage, and the judgements made by teachers at the end of a key stage. Everyday assessment is an integral part of teaching and learning; it is how teachers gain knowledge of their pupils' needs, achievements and abilities. Statutory teacher assessment involves teachers using the knowledge gained from everyday assessments to make and record their judgements on pupils' overall attainment at the end of a key stage."

This clearly outlines the role of teachers in the formative and summative assessment of their pupils. On-going, or formative, assessment is essential if pupils are to increase their knowledge and understanding in geography while widening the range and complexity of the skills they use. As in any subject, teacher assessment in geography has to cover the full range from informal observations and conversations to more formal assessment activities.

The Geographical Association (GA)

brought together a group of people interested in assessment in geography, including one of the authors, to produce a book called *Assessment Works: Approaches to Assessment in Geography at Key Stages 1, 2 and 3*. This book contains some useful guidance on assessing National Curriculum geography. Figure 6.2, taken from that publication, shows clearly the place of formative and summative assessment in teaching and learning in geography. The starting point for your planning is the programmes of study; from there you move around the cycle. It cannot be stressed too much that assessment is integral to good practice in teaching and that sound curriculum planning is essential to support sound teacher assessment.

Figure 6.3 lists some crucial things to remember when assessing geography.

The Revised National Curriculum has introduced *level descriptions* to:

> *"provide a more helpful, realistic and manageable basis for making summative judgements"*
>
> (SCAA, 1995)

They are a series of statements that demonstrate the characteristics of pupils working at a particular level. They should not be applied to isolated pieces of work but to the

Figure 6.2 Assessment in the greography National Curriculum

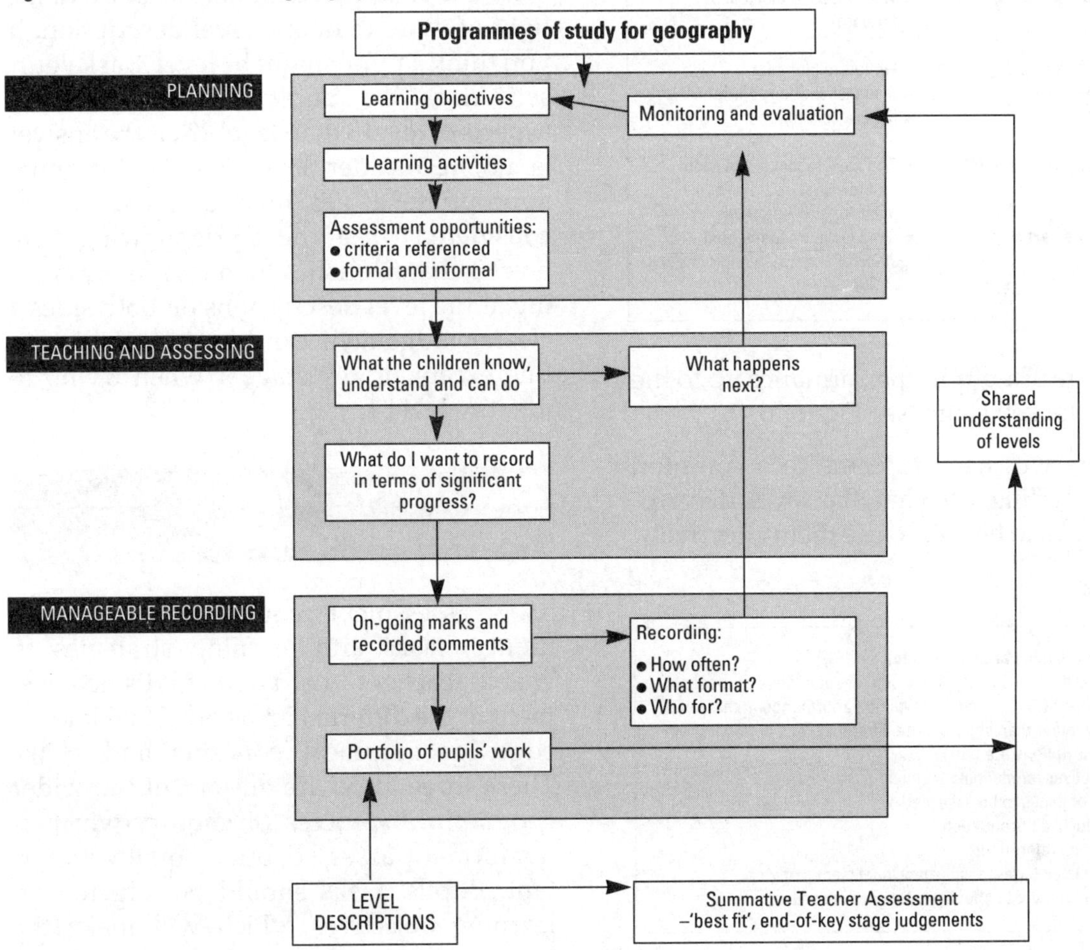

From Butt, G., *et al.*, *Assessment Works* (Geographical Association, 1995)

Figure 6.3

Teacher assessment in geography: Things to remember	
What learning outcomes do I want?	• Consult the school's long-, medium- and short-term planning • Refer to the PoS • Design the lesson learning outcomes
How will I know when and if those learning outcomes have been achieved?	• Use a variety of assessment techniques • Consult a range of evidence • Do not use the level descriptions for day to day teacher assessment • Develop your own judgements, ideas of progression
What significant achievement has taken place in understanding of place, human or physical processes or patterns and environmental geography?	• Record 'significant progress' in teacher's notebook or individual record • What do I need to go over again, consolidate? • Who needs to be moved forward?
Do I need to record anything?	• For future planning • In my teacher's records • For the next teacher
What do the pupils need to do next? *or* What is the next step?	• Do I have to make provision for individual pupils or small groups? • Can I move on to the next stage of understanding or new skills?

full range of a pupil's performance up to the end of the key stage (see Figure 6.4).

Teachers will need to come to a common understanding of what the level descriptions mean to be able to use them effectively.

Figure 6.4

Level descriptions are intended to:

- Be used at the end of a key stage for summative purposes
- Be 'best fit' descriptions of typical performance at that level
- Unify rather than separate the different areas of geography.

Level descriptions are not:

- Objectives for planning teaching
- Lists of things to be assessed
- Helpful in differentiation
- Criterion referenced
- To be used for assessing individual pieces of work
- To be turned into checklists for pupil records.

Within the level descriptions, some geographical terms are used that non-specialists might find confusing, for example 'range' and 'human processes'. These same terms are used in the PoS so any discussion and in-service on the level descriptions can only support further teachers' understanding of geography. The two level descriptions shown in Figure 6.5 were annotated by the team of writers involved in *Assessment Works* (GA, 1995) and illustrate what some of the terms mean.

Level descriptions should be used as a 'best fit' assessment system. This means that to achieve a level, pupils do not need to achieve all aspects of the level description. It is easier to make judgements if teachers look at a pupil's overall performance in the light of more than one level description. If you think a pupil might be level 3, ask yourself 'does the pupil demonstrate more aspects of level 3 than level 2?'. If the answer is no, reconsider level 2 – is this more appropriate? If yes, then level 4 should be considered to see if he/she is showing more level 3 characteristics than level 4. By looking at the level descriptions on both sides a clearer judgement can be reached. This is fine except in key stage 1 when trying to identify level 1.

Creating opportunities for teacher assessment

'Good assessment' requires a range of techniques, tasks and learning strategies to enable teachers to judge a child's achievement in the different areas of factual knowledge, geographical concepts and skills. These three areas are all part of the wider geographical process of enquiry which is not formally assessed, but is equally important. Pupils' tasks should be targeted on learning objectives which will make the

Figure 6.5

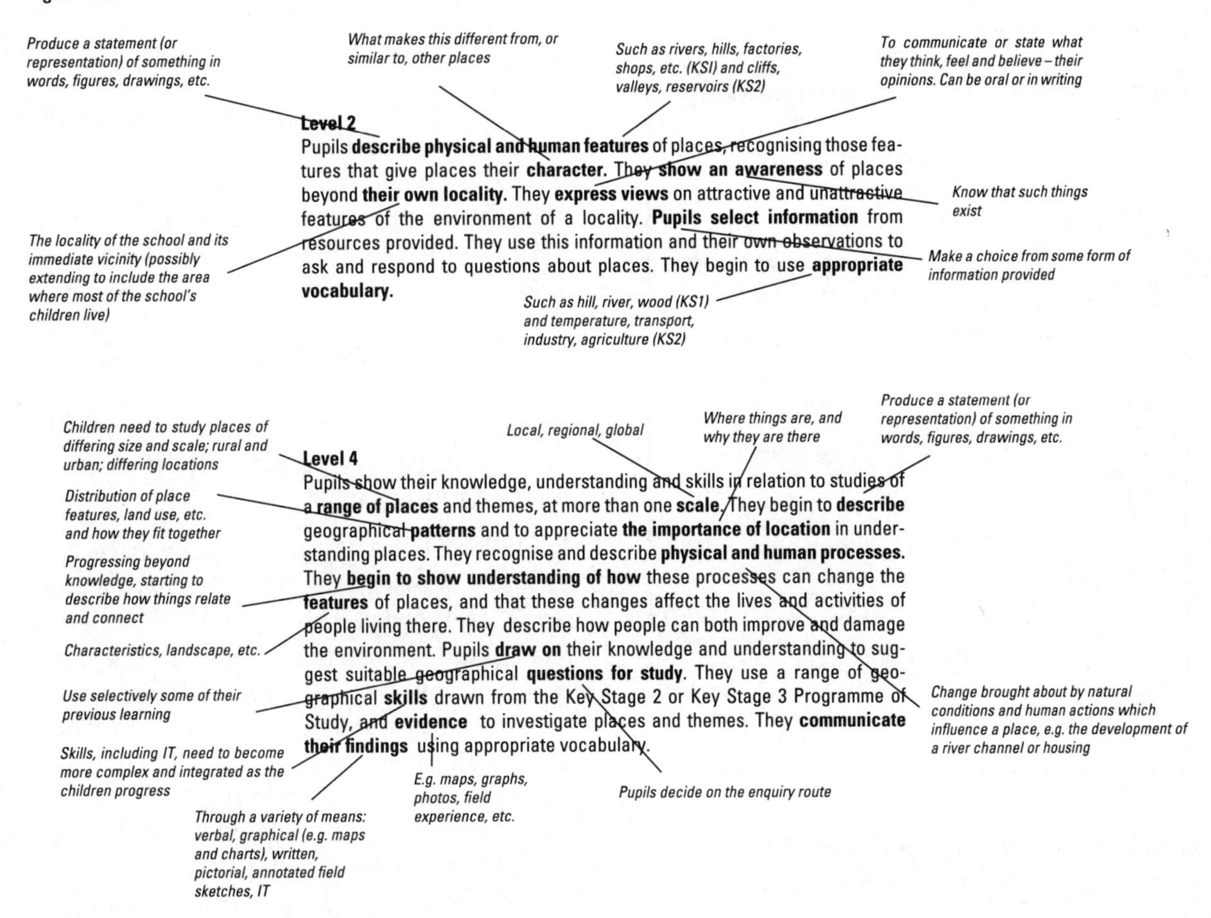

From Butt, G., *et al.*, *Assessment Works* (Geographical Association, 1995)

outcomes or evidence valid, meaningful and easier to access. The types of evidence that can be used to show pupils' achievement can be split into teacher and pupil evidence. The types of pupil evidence are shown in Figure 6.6.

Apart from photographs, teacher evidence is mainly in the form of:

- Observations recorded in notes, records or annotation slips
- Pupil records
- Class records
- Notes in notebook.

It is best to develop a range of teaching and learning strategies; this will give the pupils a variety of ways in which to demonstrate their achievement. Some children will achieve at a higher level if asked to present their evidence in a certain way; for example a child with poor coordination might perform better verbally than graphically. The chart 'Creating opportunities for teacher assessment in primary geography' (Figure 6.7) contains some ideas of children's activities that can be built into units of work and used for assessment purposes. This is by no means a definitive list but can be used as a starting point to develop your own ideas. Figure 6.8 is an assessment check grid. It enables you to check how many different types of evidence you use now or have planned into your geography units of

Figure 6.6

Pupil evidence	
Written evidence	**Graphical evidence**
reports notes diaries questionnaires stories essays newspaper articles short-answer questions multiple choice questions cloze procedure	maps drawings diagrams graphs print-out labelled photographs landscape sketches
Products	**Oral evidence**
models artefacts	questioning discussion interviews sequencing explanation hypothesising describing evaluating role play pupil presentations tapes video recordings debates

work. This will help you to maintain a balance of different tasks over the key stage.

Good formative assessment practice

Once you have made your assessment and feel that the children are ready to progress in developing their geographical understanding, knowledge and / or skill, the problem usually faced by teachers is 'what is the next step?'

Within the National Curriculum Geography Order there are four main areas of geographical understanding:

- Knowledge and understanding of places
- Knowledge and understanding of geographical processes and patterns
- Knowledge and understanding of environmental issues
- Geographical enquiry and skills.

These are developed within the framework of the places and themes within the appropriate PoS. If the PoS are considered alongside the level descriptions, broad lines of progression can be seen. Figure 6.9 is by Alan Waters, the Geography Inspector in Cumbria who has been working on progression within the new Order. Obviously a chart like this does not answer all teachers' questions, but it does help to identify elements of progression and support teachers' understanding.

In 1996, SCAA published *Consistency in Teacher Assessment: Exemplification of Standards, Geography*. This material is aimed at key stage 3 teachers, but as the level descriptions cover all three key stages it will also be of interest to key stage 1 and 2 teachers. The first section discusses progression in the four main geographical areas giving examples of the main pupil characteristics at the various levels, starting from level 1 and demonstrating this with examples of pupils' work. The second section is three pupil profiles, including one at level 4 and one at level 5. This book would be helpful for teachers who are beginning to collect example portfolios and help them come to a greater understanding of progression in geography.

It must be remembered that pupils will only be able to demonstrate understanding and knowledge if:

- Teachers' planning allows them to
- The appropriate activities have been designed
- They are asked to present their thoughts and show their understanding or use their skills. For example, they need to have worked on human and physical

Figure 6.7

Opportunities for teacher assessment in primary geography

Type of activity		Geographical context
OS map	Using and interpreting maps	Story to include features you can see from a given grid reference; Postcards: stand at point X and draw what you can see. Swap cards with other pupils – where was your postcard written? Give a detailed account of a journey from points A to B by car or foot; signpost maps from given grid references.
Atlas	Make your own postcard	Postcards to show which country you are in. Give at least three clues in your picture and writing.
Map and plan	Giving directions to other children	Can they follow instructions? Where did they end up? Construct maps for younger children. Do they need a key? How much information to put in?
	Listen to instructions	Draw a map of X's route to school.
	Types of maps	What sort of map would you need to ...? Who would use ... type of map?
Factual recall	Normal classroom writing	Write two pieces of work, one at the start and one at the end of the unit of work about a place/issue; make up a crossword, word search to include some clues and facts on a given topic. Devise a test for your friends and construct an answer sheet.
Vocabulary	Explanations of vocabulary and concepts	Write a sentence/paragraph/draw a picture to illustrate x; geographical cloze procedure with or without words. Use the following geographical terms correctly, e.g. hill, reservoir.
Drama	Role play and/or assemblies	If possible these should be written by the children, e.g. a day in the life of a family in India.
Tape	Tape recording of discussion or explanation	Explaining settlement types from pictures.
Artwork	'Before and after' pictures	Drawings in different mediums, e.g. mountains and rock strata or deserts.
Treasure trails	Directional competitions	Design a trail. Follow a trail.
	Orienteering	Collect letters to make up a word.
Playground games	Directional games	Bouncing ball on a large map of unnamed countries. Move to N, S, E, etc.
Board games	Designed by the children	Cross-curricular games (geography, technology, maths, English): routes, treasure islands, collecting sets of different shop types, etc.
Computer	Directional activities Data bases	Logo for spatial awareness. Roamers and Turtles can all be used to check direction and routes. Use atlas, book to fill gaps in a data base. Add climate statistics or place names.
Technology	Project briefs	Can be made to include geography, grid references taught previously, e.g. playground, park or room plans.
Poems	Write a poem	Following a route on a map or draw a map to go with a written poem.
Video	Video discussion	For teacher moderation and consideration: pupils make video about an issue, e.g. siting a new local pub, supermarket or park.
Photographs	Take or use them	Where were these taken on the trail? Can you map them, identify them or take them at given points? Spot trails are adaptable to age levels, etc. Sort and locate photographic evidence.

Figure 6.8

Assessment strategies check grid											
Year...........											Key stage........
Evidence / Unit	oral	models/ artefacts	graphics	notes	essays	role plays	video/audio presentations	factual recall	observation	photographs	other

processes, then be given the opportunity to apply this knowledge, hopefully through fieldwork, to another locality or place to be able to recognise it

- Some of the activities are open-ended, to give them the opportunity to show the depth of their understanding and knowledge, for example describing how people can damage or improve their environment.

Remember good practice in assessment, either formative or summative, allows for the unexpected as well as the intended outcomes.

When planning units of work, you need to identify assessment opportunities. Using the appropriate column, these may be informal or formal, possibly noted by you, and used for planning subsequent differentiation activities or recorded on the child's record. There might be key points in a unit of work which allows a child to demonstrate significant progress in their level of geographical understanding. A more formal task may have to be designed by the teacher to give evidence of understanding. Teachers need to track their pupils through these to be able to report on a child's progress.

Plannning identifies key points of learning and incorporates assessments either formally or informally at the right stage.

Record keeping

When considering record keeping these questions spring to mind:

- Who is it for?
- What do I really need to record?
- Who is it to be done by?
- How do I use the records to inform my teaching?

Figure 6.9

Progression in National Curriculum geography through programmes of study and level descriptions

Level	Geographical and enquiry skills	Knowledge and understanding Places	Knowledge and understanding Geographical patterns Physical and human processes	Knowledge and understanding Environmental issues
1	• Make observations • Express their views • Use resources provided and own observations to respond to questions	• Recognise and make observations about features of places, physical and human	• Recognise and make observations about physical and human features	• Express their views on features of the environment of a locality that they find attractive or unattractive
2	• Describe features • Express views • Select information from sources provided • Use information to ask and respond to questions • Begin to use appropriate vocabulary	• Describe physical and human features of places • Recognise features that give places their character • Show awareness of places beyond their own locality	• Describe physical and human features • Begin to use appropriate vocabulary	• Express views on attractive and unattractive features of the environment of a locality
3	• Describe and make comparisons • Use skills and sources of evidence to respond to a range of geographical questions	• Show awareness that different places may have both similar and different characteristics • Offer reasons for some of their observations and judgements about places	• Describe and make comparisons between physical and human features of different localities • Offer explanations for the locations of some of these features	• Make comparisons between physical and human features of different localities • Offer reasons for some of their observations and judgements about places
4	• Describe patterns • Suggest suitable questions for geographical study • Use a range of geographical skills and evidence to investigate places and themes • Communicate findings using appropriate vocabulary	• Appreciate the importance of location in understanding places • Show understanding of how physical and human processes can change features of place, and how changes can affect lives and activities of people living there	• Begin to describe geographical patterns and to appreciate the importance of location • Recognise and describe physical and human processes • Show understanding of how processes can change the features of places	• Describe how people can both improve and damage the environment • Show understanding of how processes can change the features of places
5	• Describe patterns and processes (range of physical and human) • Identify relevant geographical questions • Select and use appropriate skills and evidence to investigate places and themes • Present findings both graphically and in writing	• Describe how processes can lead to similarities and differences between places • Describe ways in which places are linked through movements of goods and people • Investigate places and reach plausible conclusions	• Describe and begin to offer explanations for geographical patterns and for a range of physical and human processes • Draw on knowledge and understanding to investigate themes and to reach plausible conclusions	• Offer explanations for ways in which human activities affect the environment and recognise that people attempt to manage and improve environments

Source: Alan Waters, Geography Inspector, Cumbria

- When is it completed?
- Is it creating a duplicate record?
- What do I record it on?

In *Geography: A Review of Inspection Findings 1993–4*, OFSTED stated:

> *"Where records were kept they tended to reflect the teacher's coverage of the subject rather than the pupils' achievements in it. Assessment had little influence on planning other work."*

These types of practice in record keeping obviously do not help identify pupil progress and most certainly do not support the teacher in the planning, learning, assessing cycle.

Good records support your teaching and planning; they should supplement your knowledge of the child. The Dearing Report of 1993 said:

> *"If record systems do not provide a significant contribution to teaching and learning, there is little point in maintaining them."*

This statement is very true; good records are a tool, poor records are a chore so care should be taken to get the correct balance between informing teaching and supporting planning and taking up valuable time.

Besides being accurate and up-to-date, any record system for the Revised geography curriculum should:

- Be on-going alongside the teaching, learning and formative assessment
- Be used to inform future planning
- Help individual pupils progress by supporting differentiation
- Support the writing of yearly reports
- Contribute to the end of key stage teacher judgements
- Be passed on to the next teacher or school if appropriate
- Be appropriate for the purpose you intend
- Be useful, easy to keep and manageable.

Decisions have to be made on whether to keep individual, group or class records. Some schools have decided to keep records on groups of children's achievement in the short term or informally as teacher records, or for the duration of a unit of work, using these to support planning for differentiation during the unit of work, then inform the planning of the next area of work and subsequently feed into the pupils' individual records. This system gives flexibility and is more manageable as a working document during the teaching stage. It supports the short-term planning while tracking individuals' progress. This type of record might look at the class in three main groups:

- Those with full understanding who have completed extension activities
- Those with understanding
- Those who need further support to understand the key ideas.

Significant achievement of individual pupils can also be recorded here. This type of record can be closely linked to the learning outcomes the teacher has decided upon in their medium- and short-term planning. This makes the task of assessing the pupils' progress easier and reinforces the idea that clear learning outcomes support good teaching and formative assessment, which in turn feeds back into planning the next stage of learning. On most types of record, a dated, informed comment, however short, by the teacher is more helpful than a tick. Records in geography must move away from keeping a record of coverage and attitude, they must aid individual progression and support reporting on progress. Coverage is often recorded elsewhere in the long and medium planning documents. Annotation of these records often helps the next teacher if for some reason a locality was not covered or another aspect not taught.

Portfolios

A portfolio is a collection of pupils' work, either one pupil or different pupils, that demonstrates characteristics of pupils' performance at a particular level. Portfolios can be developed through staff INSET or collected by the coordinator. The material is illustrative of standards and if produced through INSET promotes a common understanding in geography through the school. These exemplars illustrate the standards that a school teaches to and help teachers to check their judgements. The work is often annotated as to how it was done. To help schools develop a common understanding of progression in geography and what the level descriptions mean is important to aid individual pupil's progression. They are also useful when new members of staff come to the school (see Figure 6.10). Portfolios can also include photographs of wall displays or children using fieldwork skills.

Beside examples of written or drawn work, portfolios should contain a wide range of different materials:

- A range of pupils' work, showing examples from pupils of different ages and abilities
- Photographs, videotapes, tapes, etc.
- Teachers' records or notes of ephemeral evidence
- Pupils' comments.

Portfolios should be kept up-to-date and reflect the geography units being taught in the school. They should always be kept to a managable size. If you are thinking of developing a geography portfolio, there is some good guidance in Appendix 2 of the *Consistency in Teacher Assessments, Key Stages 1 to 3* (SCAA, 1995).

Figure 6.10

Portfolios can:

- Promote common standards in geography through the school
- Support improved assessment in the school
- Exemplify how 'best fit' judgements are made
- Support new colleagues
- Help the coordinator to monitor the geography in the school
- Show others, for example the governors, LEA and OFSTED, the agreed standards of work in the school.

Reporting

With the introduction of the new Orders, the DFE has produced new guidance on reporting pupils' achievements. This was contained in DFE Circular 1/95: 'Reports on Pupils' Achievements in 1994/5' and at the time of going to print the DFEE have said they see little change in these guidelines at the moment. Key stage 1 and 2 schools have to send parents at least one written report during the school year which contains information on the following:

- In the core subjects:
  - children's SATs results
  - teachers' assessment on the level of pupil attainment
- In the foundation subjects:
  A commentary on the pupil's progress in all subjects studied.

Schools do not have to state the levels the children have achieved in foundation subjects as they do in the core subjects but they do have to report on progress. This means that teachers cannot just report on coverage and attitude, the two areas that many primary reports comment on. Reporting on progress can be difficult to do unless schools use the level descriptions to give guidance on the levels the children are working at. Many schools are using the level description to keep track of pupils' progress. Because the level descriptions are wide statements to be used in a 'best fit' manner, some teachers add comments or codes, for example by adding + or – to the level, or stating two levels 3–4, possibly underlining the one nearest to the pupil's achievement. This makes the descriptor easier to use in the tracking of progress. Whether schools report these working levels in geography to parents at the end of the key stage is up to them. Obviously good record keeping techniques will be needed to give the information needed to report on progress. It is important to be able to talk about or have a notion of progression here, may be considering progress in the four key areas of geography or by using the level descriptions.

Conclusion

The major assessment dilemma is trying to balance the three issues of manageability, reliability and validity. Any assessment activity must be manageable in the classroom, usually by non-specialist teachers. It must be reliable in ensuring children's progress and development, and it must have validity for the audiences it seeks to inform. This balancing act does become easier with an improved understanding of the elements of progression within geography.

Our priority as teachers is to facilitate pupils' learning through good planning and sound organisation.

Formative assessment supports this and is an integral part of the learning and assessment cycle. Developing techniques to assess pupils' geographical understanding, possibly using the four areas of geographical knowledge mentioned earlier, is an important area for teachers to develop as it will help in planning the next steps in a pupil's geographical development. This is the formative element of teacher assessment and should not be overtaken by summative assessment which is not developmental, summarises what pupils know, understand and can do at a given moment of time and for very different audiences.

7

EARLY YEARS GEOGRAPHY

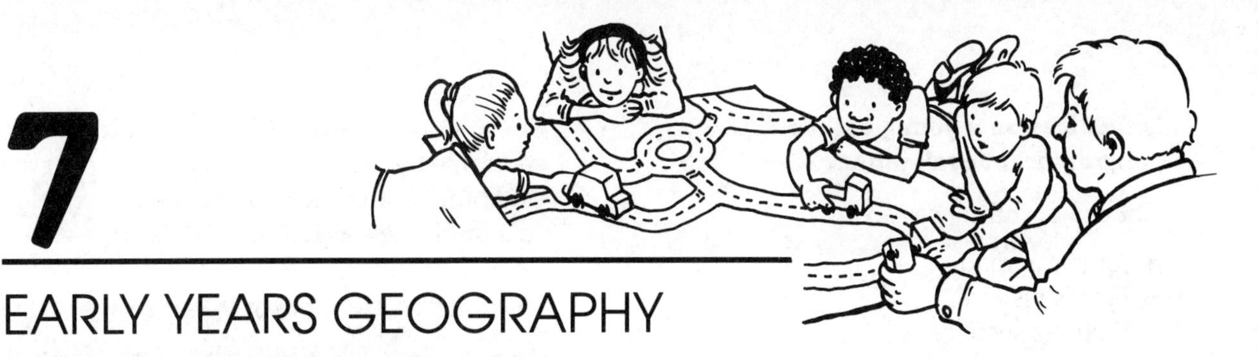

What does early years mean?

For the purposes of this chapter we are taking the definition of 'early years' to refer to nursery and reception children. This decision is supported by OFSTED:

> "*We have adopted the 'early years' as standard terminology for the 'under fives', 'pre-key stage 1', 'nursery' or 'children below statutory school age'.*"

OFSTED, *Inspection Issues and the Early Years: A Consultative Paper* (HMSO, 1995)

We have chosen to use the above definition as this seems to be the clearest one at the moment, with the following definition marking the divide between early years and key stage 1:

> "*In revising the programmes of study for key stage 1, care has been taken to ensure that they can be taught and assessed within a period of six terms.*"

SCAA does give the option, however, that:

> "*Schools may, however choose to spread their teaching of the key stage 1 National Curriculum over a longer period and thus cover aspects of key stage 1 with children in reception classes if they judge this appropriate to the needs and stage of development of the children.*"

SCAA, *Planning the Curriculum at Key Stages 1 and 2* (SCAA, 1995)

Where do we start?

The starting point has to be the child's own experience. When children first come to nursery or school they all have their own personal geographies, which encompass spatial awareness and the relationships between places, people and things that they have developed through their first years. As teachers, we have to help children develop and widen their understanding of the links between places, different environments, widening their knowledge and understanding. Figure 7.1 shows the many influences on young children's lives. It is based on a diagram by Goody first published in 1971.

Figure 7.1

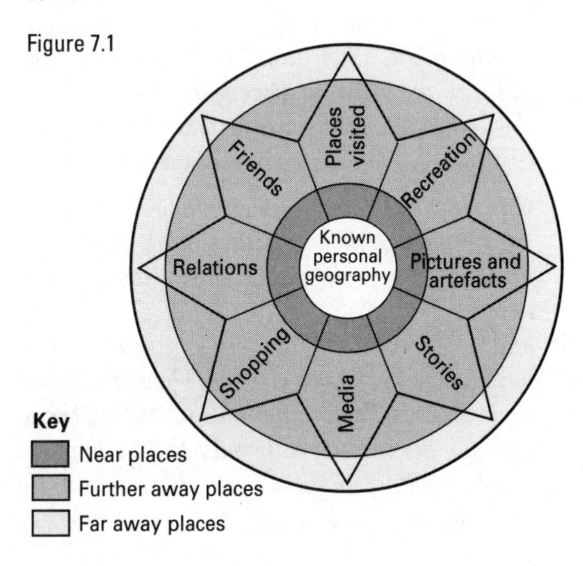

Influences on a young child's geographical development

The major influences are:

- Journeys made
- Places visited
- Pictures seen
- Places talked about.

Geographical development is a natural part of a child's development. It helps them to make sense of the world around them and their place in it.

The SCAA pre-school consultation document (1995) gives six areas for education in the early years:

- Personal and social development
- Language and literacy
- Mathematics
- Knowledge and understanding of the world
- Physical development
- Creativity.

The area 'Knowledge and understanding of the world' has obvious links with geography. They also state the desirable outcomes in each area. Those for knowledge and understanding of the world are:

> "*The outcomes relate to children's developing knowledge and understanding of their environment, other people and features of the natural and man-made world. They include the development of skills for later learning in history, geography, science and technology.*
>
> *Children talk about where they live, their environment, their families and past and present events in their own lives. They explore and recognise features of the natural and made world and observe similarities, differences, patterns and change. They talk about their observations, sometimes recording them and ask questions about why things happen and how things work...*"

SCAA, *Planning the Curriculum at Key Stages 1 and 2* (SCAA, 1995)

So a good starting point would be to reinforce the learning needed to support the outcomes stated here. Children are entitled to a quality experience.

What are the key geographical elements

The key elements of geography in the early years can be divided into the same areas as in key stage 1, with slightly different emphasis as Figure 7.2 shows.

Figure 7.2

Key elements	Emphasis in early years
Places	Features of the local area Awareness of other localities
Themes:	
1 Human	Use of buildings, e.g. shops, school People's work, e.g. doctor, postman
2 Physical	Landscape features, e.g. hill, slope, pond Weather, e.g. rain, clouds
3 Environmental	Opinions on the environment Caring for the environment
Skills:	
1 Enquiry	Asking questions: Where do puddles go? What do you do?
2 Observation	I saw a post box
3 Directions	Follow and give instructions, e.g up, turn, forward
4 Plan view	Shapes, pre-plan view activities, e.g.shape printing
5 Working outside	Environmental awareness
6 Communicating information	Describing the weather or a route
Spatial perception	Children play with and through spaces to develop their spatial perceptions. They need to explore, move around in and change their environments to aid their understanding of the many different environments they will encounter
Attitudes and values	It is commonly acknowledged that positive attitudes and values to different races, cultures and environments are easier to develop in the early years. This is a very important area for teachers to be aware of.

It is not intended that these key elements of geography be taught as a discrete subject to this age group. Geographical understanding forms part of a child's experiences of the world. So in planning a curriculum for this age group, we need to take account of where the children are, building on familiar activities and places.

How do we do it?

Work should be based on children's own personal geographies, developing and widening them, encouraging children to look at the links in their fragmented experiences. This obviously has to be planned for. Teachers need to be aware of the geographical potential of a situation when planning activities. The intended outcomes from structured play must be clear in teachers' minds and planning, then these play activities can support geographical understanding. To develop the geographical side of the activity appropriate vocabulary will need to be used.

In her book, *Geography Starts Here!*, Angela Milner states (page 12):

> "*The most effective planning will include activities based upon:*
> *the children's own personal geographies*
> *helping children to establish their own sense of identity*
> *encouraging children to search for patterns and provide their own explanations*
> *the children acting as travellers, explorers and investigators.*"

These seem to be good starting points. The authors feel that geographical activities should be based on:

- Children's own personal geographies
- Children's own varying travelling experiences
- Children's exploration of different environments
- Children's enjoyment of investigations
- Children being helped to establish their own sense of identity
- Children being encouraged to look for patterns and explanations.

As with all early years work, these activities should be developed through:

- Practical play
- Building on the children's own experience
- Observation and description by the child and the teacher
- Setting up situations for the children to investigate
- At all times using and encouraging the children to use the correct language and vocabulary.

Helping the children to develop and use the correct language and vocabulary is crucial as without it they will be unable to develop their thinking and move forward. It is impossible to stress enough the vital role played by the teachers' use of the correct language and good open questioning techniques to support the children in their enquiries, for example:

- What did you walk past when you went to hang up your coat?
- Can you show me where you went on our model of the classroom?

In the early years it is crucial to develop the children's observational and vocabulary skills. They need to identify what is around them to support their developing understanding. They also need to experience contrasting environments both rural and urban. They need to understand the effect of weather on their daily lives to begin to appreciate the impact of climate on people living in different countries. Television and photographs can help in the experiencing of different environments, but real experience if possible is preferable.

What do we need?

Most of the resources needed are commonly to be found in early years classes. It is being aware of their geographical potential that is important. Well-selected resources can support and develop young children's geographical understanding. Resources similar to those listed in Figure 7.3 can be found in most nursery and reception classes. They can all be used to develop geographical understanding and help children to make sense of their world. Teachers must be aware of clear geographical learning outcomes from a particular piece of equipment.

Figure 7.3

Useful resources for geographical activities	Possible geographical outcomes
Outdoor play area	Spatial awareness Relative location Routes Different environments
Sand and water	Landscapes Flow Plan view
Rocks and stones	Awareness of the physical world around them Geographical language Sensory development
Role-play area	Work people do Living in different environments
Dressing-up clothes	Living in different environments Different weather types
Books and atlases	Knowledge of sea and land Stimulate interest in other places
Globes and maps	Knowledge of sea and land Knowledge of other places Relative location Routes
Play mats	Routes Warning signs Directional language Relative location
Floor toys	Directional language Routes
Building bricks and construction toys	Directional language Settlements
Small world models	Different types of transport Talking about journeys Settlement features Other places
Large moving vehicles	Routes and journeys Relative location

It must not be forgotten that the most precious resource in early years geography teaching is the teacher's questioning skills that can support and extend the child's learning.

How can I use these to support geographical teaching?

The teaching strategies discussed here can be continued into key stage 1 where appropriate, either as class activities or to support differentiated activities. As children start in the reception classes at different times, the work is very often arranged as a continuous experience rather than as set topics or units of work. The activities here can be part of a continuous experience. This approach builds on good practice and most activities can continue on in a modified form into Year 1 and Year 2 either alongside planned units or as part of those units.

Structured play

This is a vital activity with this age range. Through miniature world play, children design, make and explore different environments and places – a street, a farm, a house or a garage. Rural and urban environments can be made. Teachers will be able to discuss the models with the children and through skilful questioning, develop their understanding of the types of places, use of buildings and different environments involved. Dressing-up and role-play gives them opportunities to understand other people's lives and working activities. The role play area can be used as an office, travel agents, shop, etc. Paper maché landscapes can be used with small model vehicles, houses and people to look at journeys and the beginning of map work. The sand tray and water play areas can be used to model different environments.

Geographical language

Good early years practice includes a lot of vocabulary work, extending and clarifying children's vocabulary. There is no reason why a proportion of the words should not be geographical. Words such as right, left, up, over, road, path, sea, cloud, town, hill, river, map, shop and factory are all relevant geographical vocabulary and should be explained and used. If words such as forest or cliffs appear in a story, it should be checked that the children know what the words mean, possibly following them up with pictures of different forests and beaches. Classifying and sequencing are familiar reception activities. There is no reason why sorting cannot involve sorting geographical pictures, hot and cold countries, countryside and towns, hills and flat lands, farms and shops. Postcards or pictures from magazines and travel brochures can be used. Large pictures of landscapes can be put up with three or four features clearly labelled. These become some of the words to be emphasised with the children that week. Sequencing from the largest to the smallest (a common early years practice) can be done with pictures, for example from a tree, to a wood, to a forest, or from gravel, to stones, to rocks. Talking about the weather and using the correct vocabulary while keeping a weather chart is also a good activity.

Mapping through stories

This is the idea of illustrating a story with a large map and the pupils moving the characters around the map as the story is told. Many infant books are suitable for this. These are just a few:

Rosie's Walk by Pat Hutchins
We're Going on a Bear Hunt by Michael Rosen
Dino the Dinosaur and the Volcano, Longmans Easy Start Reader
Winnie the Pooh by A.A. Milne
Have You Seen My Cat? by Eric Carle.

A large-scale map, if possible covered with clear, protective, plastic film and mounted on card can be beside the teacher. Pictures of the characters either drawn or coloured by the pupils or teacher can be moved around the map as the story progresses. In primary schools, a good design project for Year 5 or 6 pupils is to read various suitable books and in groups design and draw the relevant maps and characters for the reception class to use. This is a good way of assessing the older pupils' spatial awareness and map skills. Large-scale maps of the school can used as a base for the children to show their knowledge of the school site by putting selected photographs of site features in the correct places.

Working with globes

The new inflatable globes are very good for this age group and a variety of sizes help to begin the ideas of scale. Children of this age group are used to playing with models of things like cars, houses and trains, so why not the world? It provides a basis for the development of geographical language. Young children are fascinated by globes and there are a lot of activities they can do with them. Some are especially designed for the early years – only the sea or the land are coloured and stickers can be used if required or children can develop their knowledge of the different shapes on the globe. Questions can be varied: Where are the hot countries/cold areas of the world? Where are we? Produced in America and just beginning to be available here are the soft huggable cushion-type globes which are very 'early years friendly'.

Directional activities and routes

An essential activity in the reception class is finding your way around your new environment – the classroom, the school building and the playground. Routes to various places can be talked through, walked in groups, in pairs with or without adults until the children are familiar with their surroundings.

An extension of this activity is to have an old black or white board and paint on it a simple outline plan of the school. Whenever the class come back from the library or hall, draw the route they have used on the plan, talking it through with them. This can be further extended – when a pupil comes back from taking a message, see if they can draw their route on the plan. Guided walks, with good pertinent questioning techniques from their teachers during and afterwards, will help children to develop their sense of direction. Many directional skills can be developed through PE and dance activities. Children can direct a Roamer, or large floor toy, around prearranged routes. Long skipping ropes, garden canes or folded newspapers can be used to outline the route.

Enquiry

By their nature, early years children are inquisitive and naturally get involved in enquiry. This can be channelled into the enquiry approach. The beginnings of this approach can be laid by developing simple activities – talking about:

- Their likes and dislikes of the classroom or school grounds
- What they think it would be like to live in a hot or a cold country
- Whether they live in a house or a flat
- Can they describe their home
- How environments change.

This is by no means an exhaustive list as the possibilities to develop geographical understanding in the early years are endless.

8 MAPWORK

Geographical skills

Enquiry, mapwork, fieldwork, information technology and other, more general, skills essential to geography are crucial to good primary geography work.

In National Curriculum geography, skills provide the third dimension of the places/ themes/skills cube (Chapter 4, Figure 4.1). Once places and themes have been planned for, learning objectives which incorporate skills can be developed through key questions in units of work. It is not possible to teach geography without planning pupil activities which develop skills. However, we need to ensure that we develop all four types of skills, and that we do so with due care and attention to continuity and progression.

The importance of the enquiry process and its skills is stressed throughout this book.

Chapters 8, 9 and 11 attempt to explain the various skills and techniques which need to be planned into pupil activities as the third dimension of the work.

Tackling mapwork

Considering that for many people geography means mapwork, it was a cause for concern that HMI were not able to be positive about the teaching and learning of mapwork in the sample of schools studied pre-National Curriculum.

> *"Pupils achieved satisfactory or better standards in mapwork including the use of atlases and globes in only one quarter of the schools."*
>
> HMI, *The Teaching and Learning of History and Geography* (HMSO, 1989)

The introduction of National Curriculum geography meant that this issue was forcibly addressed. Indeed, it was addressed at the expense of places and themes, as primary teachers seized on mapwork as the part of the 1991 geography Order they could understand. By 1993, OFSTED was able to observe, when reporting on teaching and learning at key stage 2:

> *"Pupils were generally confident and able in their use of mapwork, which was the dominant feature of the work in all years but even here there was often poor continuity from one year to the next."*
>
> OFSTED, *Geography Key Stages 1, 2 and 3, Second Year 1992–3* (HMSO)

The reason for this could be that primary teachers have not had much help in identifying the sequence of skills and experiences children need to cope with maps.

Most children have a natural curiosity about maps in spite of their abstract nature. With an awareness of the continuity and progression needed to build up these map skills and experiences we should be able to direct this curiosity towards improving their learning.

The Geography Working Group advised us in their Interim Report in 1989 that, 'In map-work, maps may be created, used, compared, or analysed either in the classroom or field.' In other words, pupils need to make, use and read maps in the classroom and in fieldwork. Ideally, children should develop their mapwork in a context relating to a specific place or theme – not as a mapwork topic or lesson. They should be making, using or reading a map because the need to do so has been generated by a wider activity relating to place and theme.

With these issues in mind this chapter offers an analysis of the detailed skills and issues of mapwork in the primary years. Children at primary level need to be introduced to a wide range of different types of map making, as follows.

Making maps from first-hand observation

Freehand or sketch maps When we draw freehand or sketch maps we are trying to fit some real, larger area which we see around us – be it desk top, flowerbed or recreation ground – on to a small sheet of paper. Both children and adults can be fearful of this activity because they feel they have to get it right; but they do not have to. Because it is a sketch map not a scale map, it can only ever show relative position – where the tree is in relation to a fence and the classroom. A way of introducing this activity could be to give each child a paper with the title and one important feature of the map already in place. This gives some security for the child. Later work can start with a blank page.

Mental or cognitive maps We carry the information we need to sketch this kind of map in our heads as we cannot see the whole area to be mapped. Children may make a mental map of a route around school or of their route to school from home. When we direct a stranger to a place, we put a mental map into words, or sometimes we draw it on paper to show the way.

Imaginative maps Such maps may only exist in the drawer's mind or in descriptions in fiction, for example, children may draw:

- A map of their treasure island, and locate hidden treasure on it
- A map of Red Riding Hood's walk
- A map of a place described in a story written by themselves.

Scaled maps There is a close connection with maths here. Distances on the map may be represented in exact proportion to their real-life dimensions. It is easier for children to use a scale map from which to measure size and distance than to draw a sketch map to scale for themselves. Both activities need considerable practice, and opportunities to achieve this must be built into units of work.

Making maps from second-hand sources

Maps made from aerial photographs Tracing maps by overlaying an aerial photograph with an overhead transparency acetate or good quality tracing paper, helps develop recognition and understanding of plan form. This activity highlights the need for labelling, or for a key, because buildings and outlines so easily recognised in the photograph will become meaningless lines or shapes on the map unless appropriately annotated.

Copying maps from maps This is usually to be avoided unless there is a very good reason for it. Practice in hand–eye coordination is about the only justification, along with copying a base map or outline to create a map for a new purpose. It is better for pupils to fill in information on a blank map prepared by the teacher than to spend time free-drawing or tracing an outline.

Working with partially completed maps This activity can only come after early free-hand map-making experiences as it assumes some understanding/recognition in the children that what they are being asked to deal with is a map. Using a partially completed map assumes that children have:

- Some idea of plan form and symbols
- Some idea of relative scale
- Some ability to turn the map to the right position relative to surroundings
- Some ability to follow simple instructions to complete the map.

Remember to train your pupils to include on every map they make:

- A title
- A north arrow
- A scale, approximate or accurate.

Infants should manage a title as a minimum requirement; the other elements will be included according to a child's age, ability and experience. By the end of key stage 2, it should be possible to incorporate all of these for most maps.

Using maps

It is essential that children have the opportunity to use maps in the field (see Glossary). They should be able to use on-site as appropriate:

- Their own maps
- Maps produced by their teacher
- Their peers' maps
- Maps made by older children
- Commercially produced maps, including play maps and plastic floor maps.

Using maps on-site and in the classroom develops map-reading skills. A map communicates information to its reader via signs and symbols. Pupils may need to read just one or two sets of information from a map – the number and location of telephone boxes or the height of the land. Older juniors may be able to begin to describe the kind of landscape they will expect to see when they visit a place on an OS 1:50 000 scale map they are looking at. The key question 'What will the area look like when we visit it?' will prompt them to use their map-reading skills. They will be able to confirm or deny their predictions when they make their visit.

Types of maps for the primary school

Ordnance Survey maps and plans The Statutory Order has made these common sense for geographical work. Large-scale maps of the local area are of paramount importance. The checklist for skills and equipment in the geography Statutory Orders (see Figure 8.1) shows this, as does the resources audit check list included in Appendices A3 and A4.

Many LEAs have OS copyright licence certificates which allow schools to obtain relevant sections of OS maps for school use as long as the photocopier in an educational establishment is used. Local authority planning offices can often provide sections of maps or a single photocopy on the same basis. Some LEA teachers' centres or subject resource centres hold copies of OS maps for loan or limited photocopy.

If a school cannot access OS map copies in this way, then good local bookshops stock the OS 1:50 000 and Pathfinder 1:25 000

Figure 8.1

Checklist for skills and equipment listed and implied in Statutory Orders				
	Mapwork – making, reading and using maps		**Fieldwork**	
	Skills	**Equipment**	**Skills**	**Equipment**
Key stage 1 Enquiry process ongoing ↓	Identification of features on photographs and pictures	Photos, pictures, postcards - ground level and oblique aerial	Observation, follow directions, follow a route using a plan Record weather observations Data collection	Plans – teacher-made or pupil-made Notepad or micro tape-recorder
	Draw round 3D objects	Maths shapes, models, wooden tracks and toy buildings		
Vocabulary development practice	Using large scale maps, make plans and maps	OS 1: 1250 or 1: 2500 maps - photo-copied extracts blown up Teacher drawn maps	Make a map of a short route experienced	Notepad
	Use letter and number coordinates	Computer and software		
	Identify land and sea on globes	Globe	Use eight compass points and follow directions	Direction compass
Key stage 2 Enquiry process ongoing ↓	Use letter and number coordinates to locate features on a map. Identify features on aerial photographs	Aerial photographs	Data collection for weather measurement	Pupil-made, parent-made or commercial rain gauge, thermometers, wind vane Maximum–minimum thermometers, barometer, anemometer (wind speed gauge)
	Making maps Use four-figure coordinates	OS 1: 50 000 maps or any other suitable map		
	Measure straight line distances on maps	Cotton, ruler, 100 cm tapes	Record evidence: questionnaire making, listening, note taking	Notepad or micro tape-recorder Clipboard
	Identify features on related map and relate map and aerial photographs	e.g. OS 1: 10 000 map and 1: 10 000 scale aerial photograph	Taking photographs, sketching and labelling sketches	Camera
Vocabulary development practice	Use index and contents page in an atlas Use latitude and longitude	Atlases	Height measurement Distance measurement Space mapping (quadrats) Profile or transect drawing	Clinometer Surveyor's tape, trundle wheel, metre rules, hoops, quadrat frames, Canes or poles, surveyor's tape, trundle wheel or metre rules
	Make a key Devise own symbols Use symbols and a key Draw sketch maps			
	Orientate a map on site using land-marks and/or a direction compass Use six-figure grid references Interpret relief maps. Follow a route on OS maps Describe routes on maps Use thematic maps	Directional compass 1: 50 000 maps or map extracts Atlas maps OS 1: 25 000 maps OS 1: 50 000 maps Atlas maps showing rainfall, population, etc.	Measure width/girth, e.g. of trees Measure rate of flow, e.g. of a stream	Callipers Stop watch
	Show that a globe can be drawn flat	Orange peel!		

series. The larger-scale maps have to be obtained from specialist suppliers. Some of them will supply parchment-quality maps, others digitally produced ones which are like high-quality photocopies on thinner paper – cheaper but less robust. Multiple-copy sets of OS 1:50 000 map extracts are held by secondary schools because of each year's GCSE exams. You could try asking the geography departments of your local secondary schools for some unwanted sets. They are unlikely to be of the local area. Colour copying of OS 1:50 000 map extracts borrowed from secondary schools is possible, but cost is an issue if multiple copies are required. Borrowing the large-scale OS maps mentioned below from a secondary school geography department in order to take a photocopy is a further possibility. Remember to check that your LEA does hold a licence for copying in school. If your school is grant-maintained, you need to obtain and pay for your own licence.

Acquire the largest-scale maps first, as they are the least easy to get hold of and the most useful. Provided the LEA holds a licence, small sections can be blown up on a photocopier for practical use. Parts can be blanked out for children to fill in on-site. Finally, once safely covered with clear, tacky-backed plastic, children can trace their own sections on transparent overlay with overhead projector pens. The following OS maps will be useful:

1:1250, 1 cm to 12.5 m (old 50″ to a mile) Each map sheet covers an area 500 m by 500 m and so measures 40 cm by 40 cm. These maps are available for urban areas. All building numbers or names of houses and roads are shown.

1:2500, 1 cm to 25 m (old 25″ to a mile) One sheet covers an area of 1 km^2 and measures 40 cm by 40 cm. Often double sheets are produced – useful if your school is on the edge of two single sheets! Each double sheet covers 2 km from east to west and 1 km from north to south, measuring 80 cm by 40 cm. Individual buildings, some names of roads and large buildings, along with boundaries, field acreages, spot heights and bench-marks are shown. Rural schools will find these are the largest-scale maps available to them.

1:10 000, 1 cm to 100 m (old 6″ to a mile) One sheet covers an area 5 km by 5 km and so measures 50 cm by 50 cm. Roads, buildings or building blocks and boundaries are shown. Contour lines are shown in brown.

1:25 000, 4 cm to 1 km (old 2½″ to a mile), Pathfinder series Each map covers an area 20 km from east to west by 10 km from north to south, measuring 80 cm by 40 cm. Colour begins to be used extensively on these maps: roads in orange, contours as fine orange lines, woodland in green and streams and lakes and sea in blue. A key accompanies the map. Children sometimes have these maps at home, especially if their parents walk.

1:50 000, 2 cm to 1 km (old 1″ to a mile), Landranger series Each map covers an area 40 km by 40 km and so measures 80 cm by 80 cm. Colour is very extensively used. Children are most likely to have these maps in their homes. Although these maps are the most abstract for children, they seem to enjoy using them.

1:50 000, 2 cm to 1 km, Tourist maps Some examples of these are of Snowdonia and the North Yorkshire Moors. They can be useful for local area, regional/wider localities or contexts, or contrasting locality work, depending on the school's location.

1:50 000, 2 cm to 1 km, Project maps Each map is of a particular area, largely centred on towns with specific human features or country areas which have particular physical characteristics and are frequented by tourists. They are published in desk-top size extracts, measuring 56 cm by 45 cm. These really useful extracts can be very practical if they relate to your own local area and its wider locality, or to a contrasting locality with which you have first- or second-hand links. Pricing permits multiple-copy purchase.

Issues relating to commercial maps Teachers are often concerned that maps, however recent, are out of date and that maps are now in metric measures, yet our road distances are quoted in miles on signs and by the general public.

It is not necessary to worry that a map has inaccuracies which show it is out of date; children need to know that the landscape is constantly being altered by people. Updating the map from their observations as they go along is part of the process of using maps, and is a constructive activity for children to do.

Commercial map scales need not concern pupils until the upper junior stage, unless a pupil is especially bright and needs to understand scales earlier. The strange British practice of using metric measurements on maps and miles in travel does need to be dealt with. Children need simply to accept that we use the two different systems and begin to relate to them. Remember the useful 5 miles = 8 km relationship. The multiples of this can also be helpful: 50 miles = 80 km and 500 miles = 800 km.

In practical terms, the best way to get children to understand the issue is to ask them to think of a local distance that they are familiar with and preferably have just walked. If this distance is about a kilometre, then a mile can be related to this. One mile equals one and three-fifths of a kilometre or one kilometre and just over half as much again. If pupils in a class have travelled abroad, for example in mainland Europe or Canada, then the teacher can remind them how kilometres are used on road signs there.

Other commercial maps Pupils should have access to and be allowed to become familiar with the wide variety of other maps available – AA and RAC motoring maps, town plans, A–Z town maps, tourist trail maps and diagrammatic maps such as the London Underground ones. These may be used in locality studies or in relation to the theme of transport and transport patterns. Foreign maps – with kilometre distances, of course – may be used in the context of a locality or region in an EU country. Pupils will often have maps at home that they will be allowed to bring into school.

The elements of mapwork

Understanding maps is a very complex process. Many elements built up and brought together lead us to be able to make or read a map effectively. Practice of all kinds of preparatory concepts is needed before maps can be made. These skills are:

- Location
- Direction
- Representation
- Perspective
- Distance.

Figure 8.2 shows what each element is about in terms of mapping concepts and skills. Within each element is a progression of learning in which children need practice.

Location

There are three clear stages in location work:

Figure 8.2

Key elements of mapwork	
Location is about *Grid references*	Where? 'personal' ➤ 'relative' ➤ 'absolute'
Direction is about *Compass directions*	Which way? 'personal' ➤ 'relative' ➤ 'absolute'
Representation involves *Key*	How do I show it? 'pictorial' symbols and shapes
Perspective involves *Plan view*	Which relationships am I showing? 'pictorial/personal' ➤ spatial/abstract
Distance involves *Scale measurement*	How far? How big? Size? 'personal' ➤ 'relative' ➤ 'absolute'

1 Pupils first need to be able to locate their position or the position of particular feature on a map.
2 They must learn to locate their position relative to another feature or person when making a map or to locate the position of, say, a seat in the playground relative to that of a flowerbed.
3 Finally they must understand that absolute position uses an abstract grid to relate one object's position to another's, as in maths coordinates. This is referred to in geography as grid references. An age-old way of remembering which coordinates to take first is 'along the corridor and up the stairs'.

When dealing with grids, the progression is as shown in Figure 8.3.

Traditionally grid references or coordinates form part of paper and pencil or crayon exercises – often in maths books. More recently Information Technology (IT) has introduced another motivating way of dealing with practising grid locations through software like *Mapventure* and *Adventure Island* (see Chapter 13).

There are also many 'concrete', fun activities which infants and lower juniors can do to become familiar with grid references. They can overlay three-dimensional models and maps they have made themselves with a grid and then locate features or objects.

Model people, animals, streams and trees can be stood on card. String, coloured paper or even rulers can be used to make a simple grid. This can be lettered and numbered as appropriate.

A three-dimensional model of, say, the route Little Red Riding Hood took to get to Grandma's house or a magic island may need to be given side supports before a grid can be made across the top for children to look down on the model to locate features. Similarly, a shoe box with models in it could have a very simple grid, two squares by three, pegged over it for children to 'read' the location of objects in the model.

A large-scale map of the local area, which the children have helped to draw, laid out at their height on a low table, can have a string grid made over it in order to locate the absolute position of features. If the children understand where the grid comes from in the first place by having a hand in making it, they are far more likely to be able to read off positions.

Often the pond in the school grounds offers the opportunity to reinforce grid reference work with juniors. Some ponds are in paved areas – the paving lines offer a ready-made grid to be 'extended' across the pond with bamboo canes or metre sticks so that pupils can locate a clump of weeds or surface weed when they 'map' the pond. A variety of methods can be used to 'label' the grid – paper held down with stone, for example. The pupils will no doubt invent their own.

For a small pond surrounded by grass or tarmac, a bamboo or string grid can safely

Figure 8.3

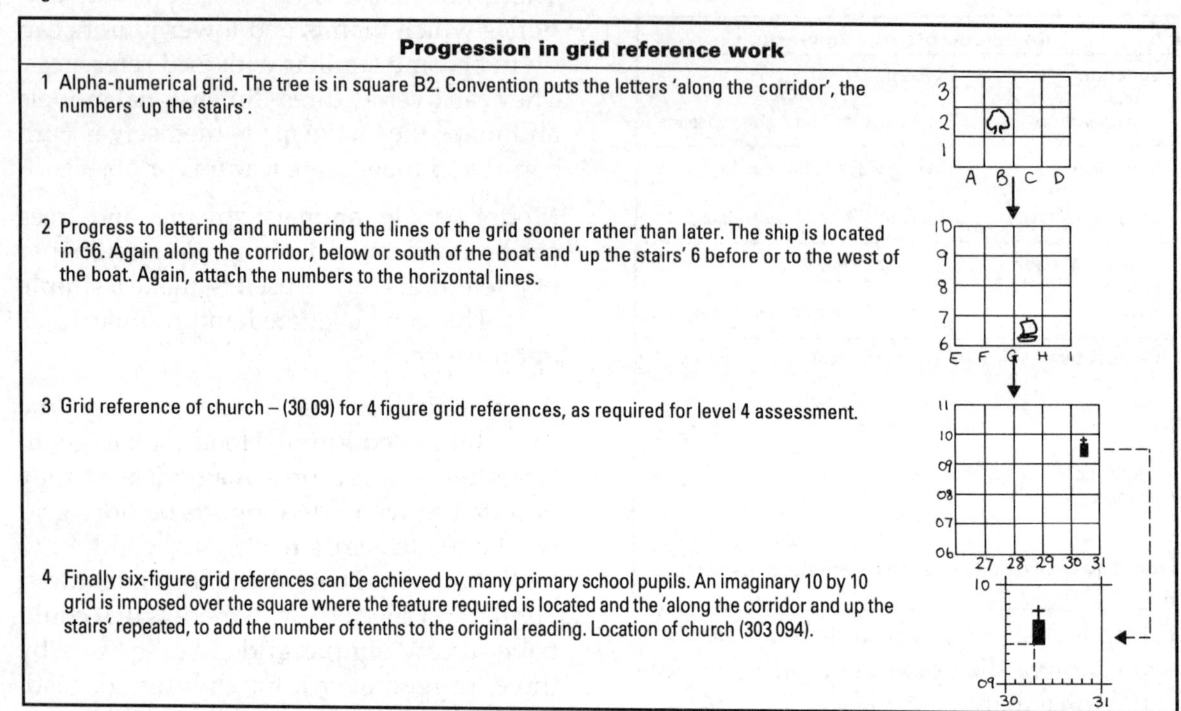

be made over it. Some ponds even have permanent metal safety grids over the top – a ready-made geographical aid!

A good self-assessment task using coordinates is for the pupils to design a ship or a house on squared paper, then swop coordinates with a friend and draw from their instructions. Do the two pictures match?

Direction

A great deal of practice in using directional vocabulary and following directions is needed before direction work can be related to maps. The concept is a difficult one, but active learning in the classroom and in fieldwork brings success. Points 1–16 suggest stages in the progression for direction work.

1 Infants need to discuss direction using simple terms – 'this way', 'that way', 'over here', 'over there', 'up', down', 'beneath', 'above', 'turn left', 'turn right', 'carrying straight on'.
2 They need to play directional games in PE and in the playground to become practised at simple direction finding and telling.
3 Games involving turns – such as quarter, half and three-quarter turns – can also be introduced and used throughout the infant and junior school years to reinforce directional concept and compass directions.
4 Then introduce shadow sticks and the link with the sun. An extension of this is to introduce compass points showing that the sun rises in an approximately eastern direction, is due south at midday Greenwich Mean Time in winter and 1.00 p.m. in British Summer Time. The sun sets roughly in the west. It is not possible to be more accurate than this in the UK – but this is sufficient. North can then be located and a direction compass repeatedly used in the school grounds. Try checking for yourself that you know

where east, south, west and north are in different locations – as long as the sun is out! Then it is easier to be more confident when helping pupils.

5 Children can chalk their own compass rose on the playground if the school does not have one already marked (see Figure 8.4).

Figure 8.4

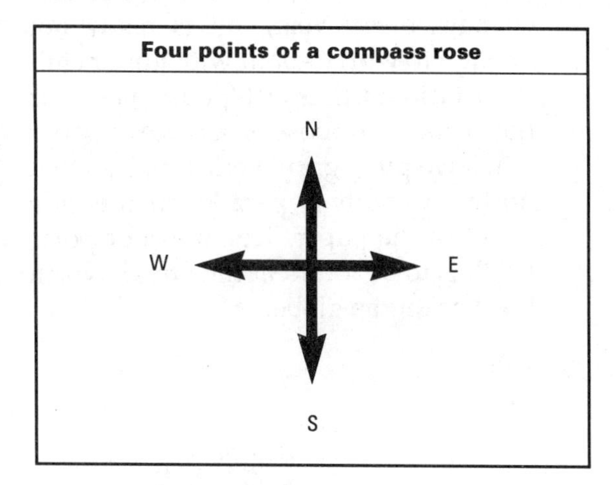

6 Now further route-finding games can be played along the lines of 'Walk five paces north, three paces east', and so on, with peers and teacher calling instructions.

7 Infants can try mapping this direction game on 2 cm square grid paper from a friend's instructions (see Figure 8.5).
All sorts of permutations on directional games of this kind, played in groups, pairs or as a whole class, reinforce the concept.

8 Control technology toys such as *Roamer* and *Turtle* are excellent for direction work where schools are lucky enough to afford them. It is possible to draw a map or a picture on an acetate and stick this over the monitor screen. Then the pupil can move the turtle from place to place on the map.

Figure 8.5

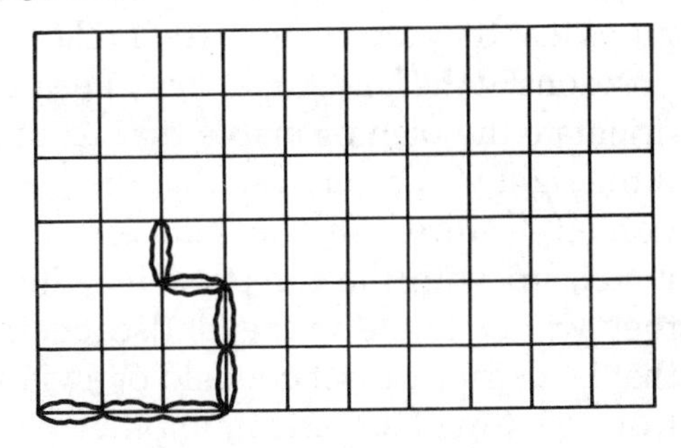

The snail trail leads:

3 paces east; 2 paces north;

1 pace west; 1 pace north.

9 A giant paper compass rose can be attached to the classroom ceiling if it is reasonably low; you can also mark one on the floor or on a surface which is low enough for infants to see and reach. The children should be involved in the correct orientation of the rose by using the direction compass to find north.

10 Every available opportunity should be taken to find north – first of all in the school grounds, then in the local area and further afield – otherwise children tend to think that 'north' only exists in the classroom. Every time the class goes out, as many direction compasses as possible should go with them, preferably one for every two or three pupils. A compass is as indispensable to geography as a pencil is to English, but it is rare for it to be used as a matter of course. It only takes a few minutes to find north and do some direction-pointing activities, whether in the school grounds with infants or on a residential field trip with older juniors. We need to remind children that magnetic compasses will be deflected from the north reading if near metal – for example a car, a metal bulldog clip on a clipboard, or a metal bar under a table.

11 Pupils should now relate plans and maps that they are using to their relative positions in the landscape. They need to orientate the plan or map – that is turn it the right way around so that the position of features on the plan or map match up with the direction in which they will be found in the landscape, be that landscape the school field or a view from the top of a castle in another part of the UK. The easiest way to do this is to locate a feature in the landscape, for example the wall of a building – find it on the map, then locate two or three more features, for example a tree or steps – work out their relative position on the map and turn the map around to match up with the general direction of the features.

12 North can be marked on the plan or map using the direction compass to find it once it has been orientated.

13 Similarly, if pupils have been given a plan with a north arrow marked on it, they can use the direction compass to relate its position accurately to the real landscape by making sure that the north arrow of the map and the compass north arrow are parallel, or that the map north arrow is lined up underneath the direction compass arrow (see Figure 8.6).

14 Once pupils can orientate large-scale maps outside the classroom, they need repeated and progressive practice in this technique, too. They should have the opportunity, every time they go out, to orientate a map on-site. The progression is to start with large-scale maps and plans and to work with increasingly smaller-scale map extracts. Plenty of practice in matching up features and then checking orientation with the compass is needed.

15 What about all the different kinds of north? Many teachers who have not specialised in geography, unless they have pursued orienteering or serious walking, do not realise that there are three types of north, as shown at the top of the 1:50 000 scale OS maps. Magnetic north is the compass needle north, actually found to the west of polar or true north, due to magnetic influences. Magnetic north is always a variable number of degrees west of polar or true north. It changes every year, but is really not enough to worry about with junior children. Grid north is just 'paper' north, or the north we impose on a piece of paper when we put a grid overlay on it so that north to south is parallel to the side edges of the paper. True north or polar north is to be found at the literal 'North Pole' as on the globe.

Figure 8.6

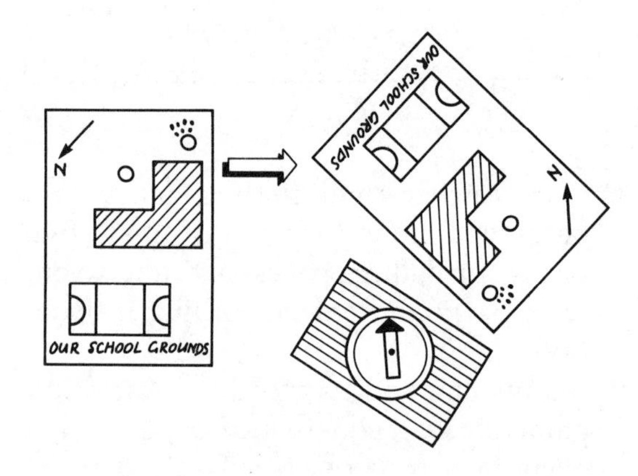

16 'Signpost' mapping. This activity can be used in context and in progression with primary children of all ages and is very useful for reinforcing directional work. Ideally, start signpost mapping with infants. Show them a real signpost in a photo or drawing and give them the concrete experience of adopting a fixed position in the classroom with a signpost pointing to other children (see

Figure 8.7). You can combine this activity with developing representational skills. How do you want children to show the location of what they were signposting

- With words?
- With pictures?
- With symbols?
- With symbols and a key?

You can also develop distance and scale measurement work. Do you want:

- Proportional length arrows or 'signposts' (that is short arrows)?
- Distances paced out and the pacings recorded along the 'signpost' line?
- The map drawing to scale?

Build in more precise directional work as you signpost map. Have pupils use the compass on site as they record their signposts and approximate directions. Alternatively, for pupils who are only able to use the four cardinal points, choose locations to signpost which represent these. 'Signpost' mapping offers many opportunities for differentiation by both task and outcome (see Figure 8.8).

Representation

As soon as the children make their own maps, they begin to explore representing life-size information on paper in their own way. They may draw pictures which they intend to replicate the feature, use the word for the feature or use a sign or symbol to show it. Infants will probably draw pictures. Side-view pictures will prevail for a long time. Children will often use word pictures and symbols side by side for years, occasionally including words, too. This is normal and acceptable child development in mapping. Figure 8.9 shows the general pattern of development in representing symbols.

Our objective is to guide children gradually towards seeing the need for a common code of symbols and a key to unlock or explain these symbols to others.

Children should have plenty of opportunity to use their own signs and symbols and talk about them to their classmates. They should not be made to feel that there is a right or wrong way of representing features on maps while they are infants or young juniors. Their development experience in plan view will encourage them to experiment with symbols, as will exposure to maps produced by teachers and commerce.

Eventually, we should lead children towards seeing the need for a commonly understood key such as Ordnance Survey maps use. Mapping in context, where the whole class has to map the same area, can be one opportunity to discuss this. When children have drawn their map, ask them to show each other their maps and get their partner to read their symbols. If their maps were to be published and sold, would thirty-six maps with thirty-six different sets of symbols and keys be helpful? How would map publishers get around this problem?

Once they understand the need for common keys and can decipher them to some extent on commercial maps, they should still have the freedom to have fun with their own symbols. Many computer programs will reinforce work on representation at different levels.

Perspective

The concept of perspective, which in mapwork is all about bringing children to an understanding of plan view, is a very difficult one. However, the basic activities which can be practised with infants to develop an understanding of plan view are

Figure 8.7

A progression in signpost mapping

Progress to graphical representation of this by getting pupils to record on paper where they are sitting in relation to pupils on their table.

SUSAN LOUISE JOHN

GARY ← ME → KIM

Extend this to objects in the room.

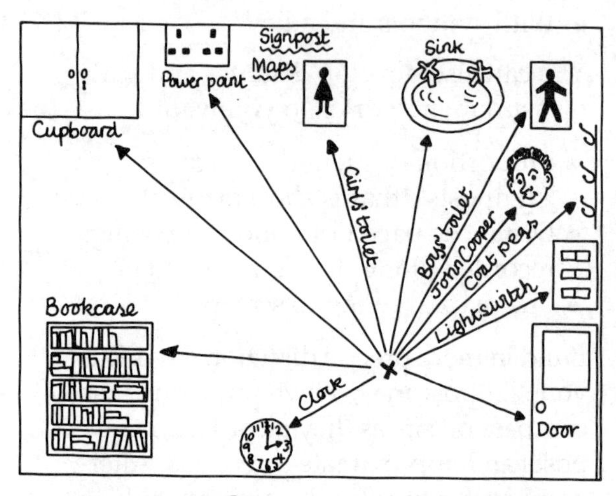

Do the activity in the school grounds, first within a limited space, e.g. the playground. Always get children to point first, then record, if they are very young.

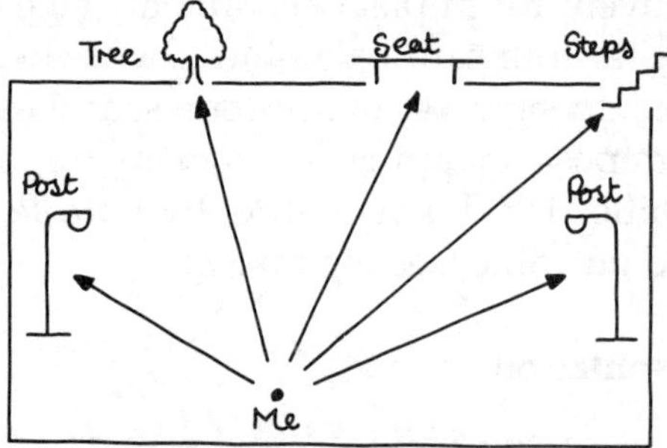

Extend the limits of the space.
Signpost map in the local area on a field visit.
Do the same activity in a contrasting area.

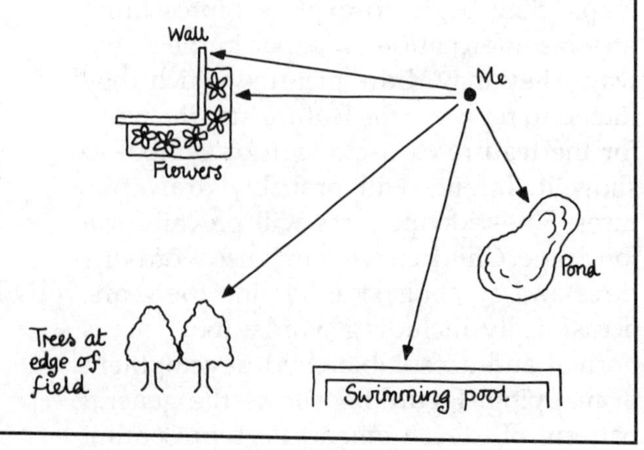

Figure 8.8

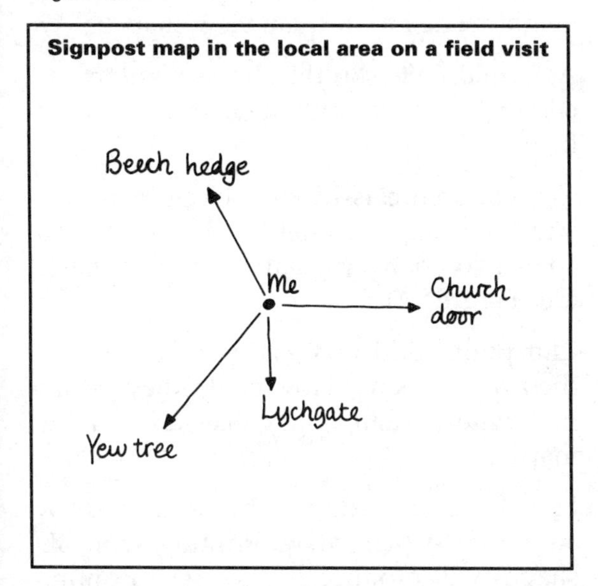

numerous and fun. Many of them involve using traditional, standard classroom materials, but with a geographical awareness. Many of the materials are standard 'maths' equipment – for example 3D shapes or 'play materials': sand, play mats, train tracks. Many materials traditionally used for art can aid the development of perspective.

Figure 8.9

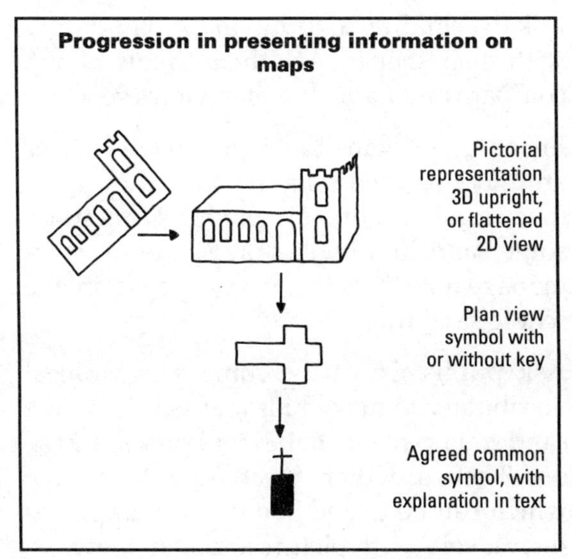

Here is a range of activities which infants and older children with special educational needs will benefit from, with the objective of developing the concept of perspective.

Using maths shapes Maths shapes have a variety of uses in a geographical context.

1 Classroom standard regular 3D maths shapes can be drawn around, creating the first map the child has ever produced. Young children may need to work in pairs, with one holding the shape and the other drawing the outline if their motor control is poor.
2 After plenty of practice with standard regular shapes, pupils can draw around more difficult and irregular shapes, for example pyramids and cones.
3 The teacher or adult helper can use the overhead projector and screen to help illustrate plan view. Help the children identify the object and sketch it, if appropriate, first from the side view or elevation. Then ensure that they look down from above to predict what they expect to see on the screen. Next project the plan view silhouette of the object onto the screen and ask the children to record and label the plan view. Beware of translucent or transparent objects such as plastic beakers, as you will get an outer and an inner silhouette for top and base – but this is in itself a discussion point.

Using everyday objects A surprising number of everyday objects can be used to support geographical activity.

1 The same activities can be done with everyday classroom and household objects, with or without the overhead projector (see Figure 8.10).
2 Using doll's house furniture with an overhead projector is a useful activity. Let the children arrange the furniture to their choice in an imaginary room shape on the

overhead projector, then let them talk out and match up the projected plan with their model room shapes.

Figure 8.10

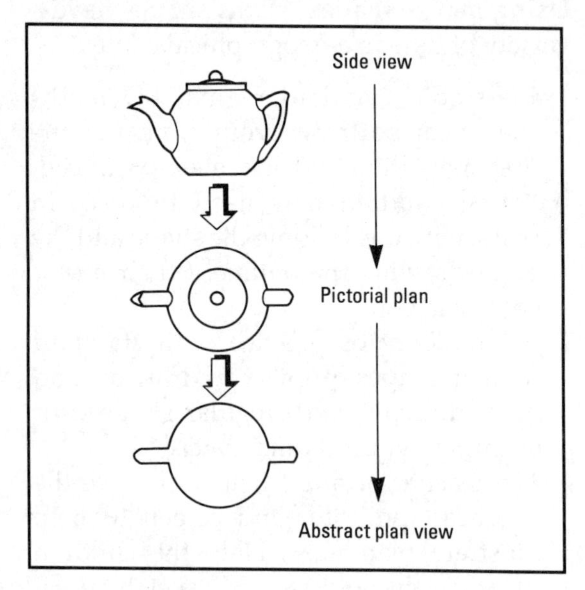

A more complex range of activities can also be developed with these everyday objects. Footprints and handprints can provide a useful starting point.

Figure 8.11

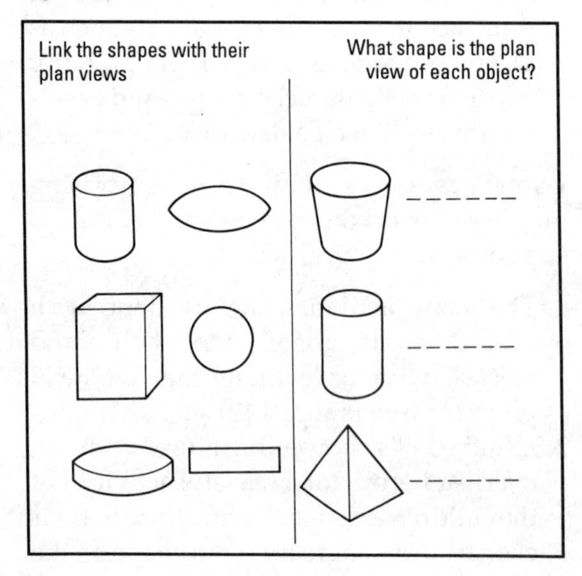

Ask children to classify a range of objects by height related to the plan view shape.

Ask children to sort the objects by function, colour or shape, and then draw around them.

Can children classify the objects by shape and then map their plan view? Is there a connection between shape and plan view? (See Figure 8.11.)

Can pupils add imaginary routes around their object maps? How could they record the route a snail might take around the map?

Printing in art can also be used to draw attention to plan view. Printing with 3D shapes will produce a plan view. Printing with flat shapes can also produce a type of map: a leaf will print a vein map if the reverse side is used to print.

Make rubbings in the local environment, for example of a manhole cover, as part of work on services is recording evidence by plan view.

Using the sand tray After free-play experience, the sand tray is ideal for developing geographical work.

Ask the children to 'print' in the damp sand with their shapes. Ask them to talk about, compare and name the plan views seen.

Encourage them to finger trace routes around 'mapped' shapes. Can they copy their sand picture on paper? Sometimes some sand in a book storage tray is more manageable than the vast expanse in the whole sand tray.

Ask pupils who have some geographical vocabulary to make hills and valleys in the sand tray, to place shapes for houses, and so on. They are then creating a landscape which can be talked about and mapped at their level – with picture and plan view.

Set criteria for the landscape after free practice – a landscape with two hills, a river, a lake, a road and a small village. Can they design the map on paper first and then create it in sand?

Using play mats Play mats *are* plan views – of roads, fields, parks, railways – although sometimes their buildings are shown pictorially rather than in plan view.

After the children have had some free-play experience with the play mat, listen to their talk when using it. Can you interject and make them aware of the appropriate vocabulary – park, playground, and so on? Can they imagine that they are an inhabitant of their playtown flying over their play mat in a helicopter? What can they see?

Using model tracks and model building blocks There are more everyday materials that can help develop plan view.

Why not ask children to link their train tracks, road pieces, etc. on large pieces of paper as a base? How could they turn their model into a map? Once they have drawn around the tracks, can they ask some friends to recreate the model on the map base? Brio® track is excellent for this.

Add imaginary 'buildings' alongside the tracks or roads, using cardboard boxes or Bauplay® blocks. Children will turn the blocks into homes, shops, buildings, and other buildings. They can be drawn around, removed and given a colour key.

Using children's own models If groups of children have made their own model village or model landscape, why not get them to draw around the model house, bridges, etc. and then remove them to show plan view?

They can then add road lines and stream lines and think how to show the plan view of a tree to complete their map of the landscape.

Placing small modelled objects in a shoe box and then mapping their plan view and location is another good activity to develop the concept of perspective.

Distance

Distance is the most difficult concept of mapwork. Although pupils do relative scale drawing every time they make a map as they are reducing real life to smaller proportions, absolute scale drawing is a difficult activity in both maths and geography. The mathematics Revised Statutory Order requires scale measurement work in key stage 3. The geography Revised Statutory Order refers to using maps and plans at a range of scales and measuring distance in KS2 PoS 3d. For both subject areas, progression in relative and comparative scale work is necessary throughout the junior school before actual scale measurement can be grasped. You will know when individual pupils in your class are ready to address accurate scale drawing.

Here is some tried and tested advice on starting scale work.

Making scale drawings Try starting with a large, regular 3-dimensional feature. How can you fit a mobile classroom shape onto paper? What about the school or local swimming pool?

How can we turn that into a diagram on paper? This can be a more stimulating alternative to the 'draw a plan of the classroom'. A simple 1 cm to 1 m scale – 1:100 – is the easiest to manage.

Fieldwork activities indicated later – such as trying to find the height of a tree or cliff – involve ground measurement and scale recording at a 1:100 scale.

Use 1 cm or 2 cm graph paper to start scale work. Either can be used on a 'one unit on the paper equals one unit on the ground' basis, or by relating the actual distance on the paper to the ground measurement.

Reading measurement scales Remember that without a scale bar or scale explanation, a scale map is meaningness.

When using large-scale OS maps or their extracts, introduce meaningful scales at some point. OS 1:1250 means that 1 cm on the map equals 12.5m on ground while OS 1:25 000 means 1 cm on the map equals 25 m on the ground. To help develop the concept of map scale, make sure children have concrete experience of what 12.5 m and 25 m feel like to walk and of what they look like.

One activity for starting scale drawing is scaling down a ruler or small box to half size, or 1:2 scale, followed by drawing a scale plan of the desk. This is, however, an isolated skill practice activity unrelated to a geographical context such as place or theme.

An enquiry activity could be used to start scale work. Faced with the question 'Can we improve the layout of our classroom?' children will need to scale map the classroom and its main features. The main features may be approximated to full metre or half metre lengths for ease, and the class grouped to produce scale plan cut-outs of the classroom, table groupings, etc. They can then move the cut-outs around on the classroom plan, think about the space between their arrangements, translating scale distance into real distance, and decide on the best layout before any classroom reorganisation. Perhaps they will discover that the current organisation is the best one!

Photographs and satellite images

There is an important link between mapwork and the use of photographs, especially oblique and aerial photographs. Both types of material can help pupils to develop the concepts of location, direction, plan view and distance as shown in the overlap diagram (Figure 8.12).

Figure 8.12

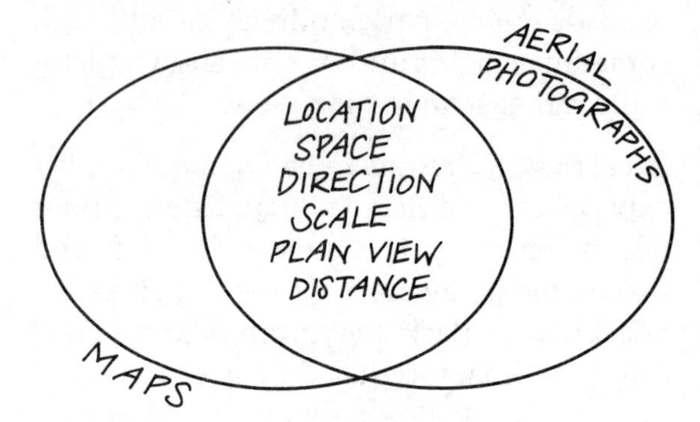

The use of photographs to develop geographical concepts and skills is new to many primary teachers. Some teachers find it useful to have the difference between maps and aerial photographs defined. An aerial photograph is an actual snapshot in time of the landscape, taken from a plane or a balloon. If it is oblique, it was taken looking down towards the ground at an angle from the plane. It is a vertical photograph if taken from directly above the scene. A map is an abstract representation of the landscape and usually includes a limited range of information, whereas an aerial photograph includes everything visible.

The geography Order programme of study for key stage 1 and 2 requires children to use aerial photographs. It is appropriate for pupils to use a range of oblique and vertical aerial photos at different scales.

Progression in the use of photographs

There is a progression in familiarising children with the use of aerial photographs so that they can identify features on them and match them to a map by the end of key stage 2. This progression is in Figure 8.13.

Figure 8.13

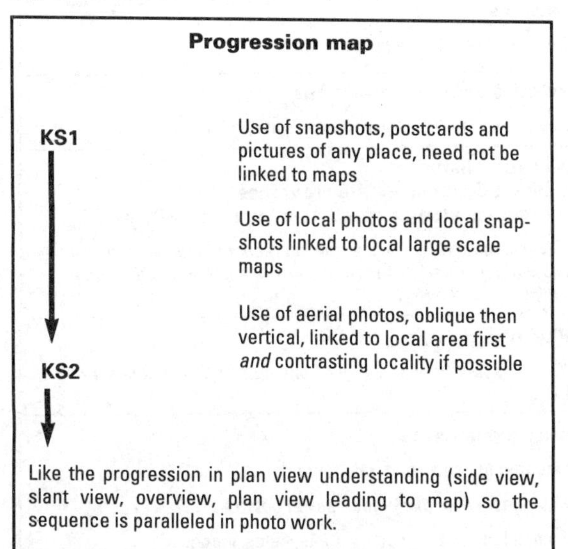

This progression goes hand in hand with the progression in viewing objects which develops plan view or perspective as already discussed (see Figure 8.14).

Figure 8.14

Viewing objects		
Actual objects		**Photos of scenes**
Side view ▼	2D or 3D	Side view photo ▼
Slant view ▼	3D	Oblique view photo ▼
Overhead view ▼	2D	Vertical aerial photo ▼
Plan view ▼	2D	Plan view photo – the more abstract, the smaller the scale
'Map' of object		

Here are some suggestions to encourage the use of photographs in geography. The use of geographical vocabulary will be developed at the same time.

Take photos around your school buildings and grounds – side view, oblique view and plan view. This need only be done once. Have them mounted for young infants to talk about and identify. Recognising familiar objects from unfamiliar angles – brick bonding, manhole covers, corners of buildings, the pond, seats – helps develop the idea of oblique and plan view.

Build up a collection of postcards. Some of them will often show oblique coastal views, oblique views taken from towers or high view-points. Children can discuss and identify geographical features freely or with your guided questions. Picture views from old calendars, etc. often have oblique views, as do illustrations from books. Local newspapers can often supply oblique and vertical aerial views, but they are usually in black and white.

Finally progress to vertical aerial photograph use. The children can use magnifying glasses to help them identify features if they want.

Figure 8.15 shows a range of activities which can be undertaken using vertical aerial photographs.

Currently the cost and availability of aerial photographs are issues for schools. A set of about six identical photographs is the most that is necessary. The class should be organised into groups, one of which can use the photographs individually or in twos. A minimum requirement is to have a set of vertical aerial photographs showing the pupils' school and its environs. If your school is twinned with a school elsewhere in the UK you may be able to acquire aerial photographs of the twin school and its locality. If the children are investigating this twin school by distance learning, then an aerial photograph is a useful information source. If they are able to visit this distant locality, then the photographs can be used before, during and after the visit as an aid to predicting what pupils expect to see and to check up on what was actually identified.

Figure 8.15

Using a vertical aerial photograph	
For ease of use the photo should be laminated! Otherwise use overhead transparency acetate overlays. Wipe-off pens essential.	**Overlay it with an alpha-numerical grid to locate features** *Key question: Where is our school?*
Use with familiar oblique or side view photos to link photo to location on aerial photo *Key question: Can you locate roughly where this photo was taken?*	**Practise compass directions** *Key question: Which way do you travel to get to ...?*
Use with corresponding OS map to match up features *Key question: Is this building really a church?*	**Measure distance** **a Direct distance as the crow flies** **b Distance along a road, river, canal or railway** *Key question: How far is it from our school to the leisure centre? Measure in 'ruler' distance or use scale according to maths ability.*
Draw a map Freehand (or same scale) by using an overlay – select out certain information according to map purpose. *Key question: Can you show why the town grew up here?*	**Plan a route** *Key question: How do we get from our school to the leisure centre?*
Identify changes on photo from personal experience *Key question: This photo was taken in 1989. Was the supermarket built then?*	**Map the land-use** Linear land use – road, rail networks *Key question: Is this place easy to get to?* Spatial land use – housing areas, fields, woods *Key question: Is there more settlement than green space in our environment?*
Talk about features *Key question: What is this?*	

Satellite imagery

The Statutory Order does not require primary children to use satellite images – images of the earth's surface taken from special satellites many miles above the surface of the earth. The satellite image represents, in a way, an extreme of the camera 'zoom lens' image. The images are taken at a very great distance away from the earth and show large areas of land. Technically satellite images are not photographs. They are responses recorded according to the amount of reflected light emitted by land, sea and man-made features. Scientifically-minded children may benefit from knowing this, but it is not essential at primary level. Nevertheless, the primary teacher who did not make use of the children's awareness of satellite images would be unwise, as primary children are constantly exposed to them in both home and school life.

In school life, the newest atlases often contain satellite images of the UK, Europe and the world as alternatives to actual maps. In home life, television presents children with satellite images of north-western Europe and the UK on the weather forecast. These satellite images help bring a better, easier understanding of the globe and of weather and climate processes, and we help children by referring to them. The complex matter of false colour interpretation and technical details is for secondary school work. The connection with atlas, globe and weather work is all that we need to deal with at primary level.

Atlas and globe work

Atlas and globe work is often seen as an activity quite unrelated to mapwork. Atlas and globe work is, however, the smallest-scale work at the end of the chain of geographical vocabulary referring to maps (see Figure 8.16).

Figure 8.16

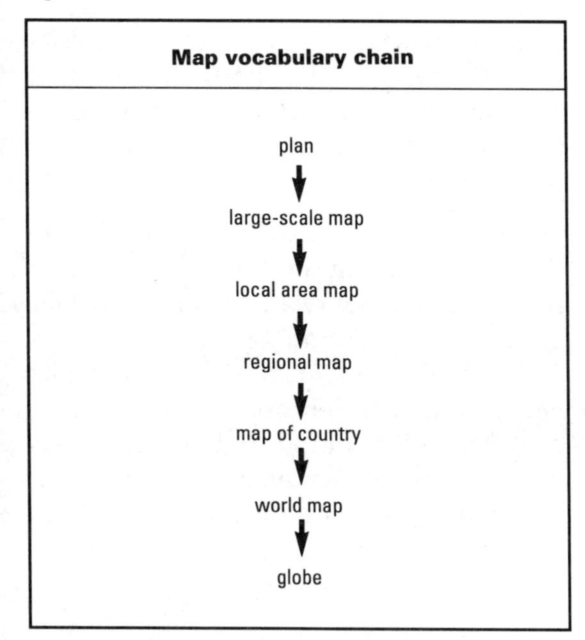

The smallest-scale representations of landscape and sea are to be found in atlases and on the globe. Children will not necessarily make the connection between large-scale maps and atlases and globes without your help.

A great deal of spatial and linear matching between maps of different scales is necessary to build up the idea of the shapes of continents and countries which, in children's minds, are adult-imposed abstract shapes. The shapes of obvious continents – North and South America, and Africa – need to be spotted on the different scale maps. India can easily look like South America to infants or lower juniors on different scale maps. A good way to help children remember a country or continent is to give them some spatial clue to help define it – India with its triangle shape, sticking out into the Indian Ocean; Italy with its boot and football island Sicily. They will be able to come up with their own animal shapes, real or imaginary, for some countries.

Political boundaries also cause immense problems for older juniors, as again there are no child-friendly criteria to explain them. It is worth getting children to trace country boundaries, or indeed any linear-type features, with their fingers first on one map, then on maps of different scales to increase their concept of spatial and linear awareness. The teacher needs to circulate around the class, and may sometimes need to pair a child who is spatially aware with a child who is having problems for this brief type of activity in the context of locating where a place is to be found or investigating its links with elsewhere.

The large-scale map of the local area can be likened to what the camera zoom lens might see if it zoomed in on a particular spot on the globe or world atlas map. Older juniors are surprised to learn that an OS 1:50 000 map they may be using covers a measurable portion of the atlas map of the UK.

Children of all ages are usually fascinated by atlas maps and the globe, even though fascination does not equal understanding. Traditionally, the use of the globe and atlas, where it occurred, took place in upper junior classes. A more comprehensive building up of understanding is likely if the globe is used constantly throughout the primary years. Each classroom needs its own. This is now a financial possibility with the introduction of inflatable globes (see Chapter 13).

Some atlas work is necessary in top infant classes and in lower juniors, too. A variety of large format, clearly presented atlases appropriate to this age range is now on the market. Individual infant atlases are now also available (see Chapter 13). Such atlases need to be clear and uncluttered, as is the Oxford Infant Atlas, and there should be sufficient in a class for a group of six children to work with, at least.

The original Statutory Order spelt out the use of globes and atlases in very specific terms through the programmes of study but related to particular levels. The Revised Statutory Order contains some very precise instructions relating to atlas and globe work, but other parts relating to this are much more vague. In key stage 1, pupils should be taught to 'use globes, maps...at a variety of scales' (KS1 PoS 3e).

This is a wider ranging instruction, replicated in the key stage 2 programme of study with a change of key wording: 'use and interpret globes, and maps...at a variety of scales' (KS2 PoS 3d).

The message here should be clear – the globe, atlas and its various maps, and indeed wall maps which can be shared by all, are the geographer's tool. Constant reference to, and use of, globes and atlas maps should be second nature in geography lessons, or parts of topics for work that extends beyond the local.

The more specific instructions for key stage 1 (KS1 PoS 3e again) are that pupils should be able to:

- Locate and name England, Scotland, Northern Ireland and Wales on the United Kingdom map
- Mark approximately on a map (presumably local, regional and/or UK) where they live
- Follow routes on globes, maps and plans
- Identity major geographical features, for example seas, rivers and cities.

This last one, however, is a very vague statement giving teachers the licence to point out and help children to identify the major features they perceive according to the context, for example seas crossed flying to Australia, as the children use the globe to try to understand the difference between land and sea.

Most infant teachers try to begin awareness of the major seas, rivers and capital cities of the UK, laying down the foundations for the locational knowledge relating to map A (the UK), which is now required by the end of key stage 2.

Specific instructions are also given for key stage 2, although in the programme of study they are not related to any particular age of key stage 2, and it is up to the teacher and coordinators to work out in which years particular locational work or atlas skills should be introduced, and how progression and reinforcement as opposed to repetition should occur. Pupils should be able to:

- Use coordinates (presumably number and letter) on maps
- Use four-figure grid references, and for more able pupils, six-figure coordinates may be taught although they are specified only in the key stage 3 PoS
- Use the contents pages and index of an atlas
- By the end of key stage 2, identify the various key rivers, capitals, physical and human features shown on the maps of the UK, Europe and the world (referred to as points of reference on maps A, B and C).

The vaguer instructions for key stage 2 also require children to:

- Measure direction and distance when using and interpreting globes, maps and plans at a variety of scales
- Follow routes in a similar way.

A minimum of opportunities must be built in for pupils to learn and reinforce these skills in each appropriate unit of work, and opportunities should be taken to develop them as topical issues arise, for example if a pupil in the class travels to Disneyland Paris or Disneyworld in Florida.

Some suggestions for starting globe work

1 Recall for the children the satellite images of the world that they often see on television, for example at the beginning of the weather forecast or on the news.
2 Talk about what the men who travelled to the moon saw when they looked back towards the Earth.
3 Have pictures of the Earth in space in the classroom.
4 Talk about how things get smaller when you go up away from the ground in an aeroplane. Maybe, if any have flown, one of your pupils had a window seat and did look down.
5 Use every opportunity, planned and spontaneous, to point out, or let children point out, first land and sea, and then countries and places on the globe.
6 Let children handle the globe, talk about it, and point out places to each other.
7 Sometimes rolling a flat atlas map into a tube can help pupils make the connection between globe, world map and atlas (atlas being a book of maps).

As the globe becomes an integral and often-used classroom resource, introduce the atlas use alongside it.

Suggestions for starting atlas work

1 Let the pupils enjoy looking at and talking about the atlas.
2 Relate places pupils have mentioned on the globe to their location on the atlas map. Pupils can reinforce this new knowledge by place-name stickers on globes, plastic floor maps and wall maps, or by writing directly on to special washable-surface floor and wall maps.
3 Always have available a large wall atlas map showing countries or relief on which to point out locations.
4 Let the children talk about and point to locations on these large wall maps.
5 Use the maps with the children to make displays. Mount pictures, postcards, drawings, cut-outs of products, and so on around the map, linked by strings to the locations of their origin. Refer to the display after it has been finished. Don't just let it merge into the 'wallscape'.
6 Get children to write about their holiday in the 'shape' of the island or country to which they travelled (see Figure 8.17). They can then trace the shape with the help of a friend, or the teacher could trace it for them, building a photocopy outline bank of Spain, France, the UK, Crete, Australia, and so on.
7 Refer to places identified on the large wall maps in the children's atlases – on world and continent maps, etc. – so that the children recognise that the same shape represents the same county or continent, even at different sizes.
8 Use every opportunity, spontaneous and planned, to point out, or let children point out, places mentioned in classwork, in the news, or by their friends.

What about the atlas knowledge required in the programmes of study?

There is a great danger that rote learning may prevail with atlas work in the primary school as a result of the Statutory Order. Apparently, the easiest way to deal with the place knowledge required for the various maps at the end of the programmes of study is to get the pupils to practise learning and drawing them, and to keep testing the memory outcome. Some schools have

Figure 8.17

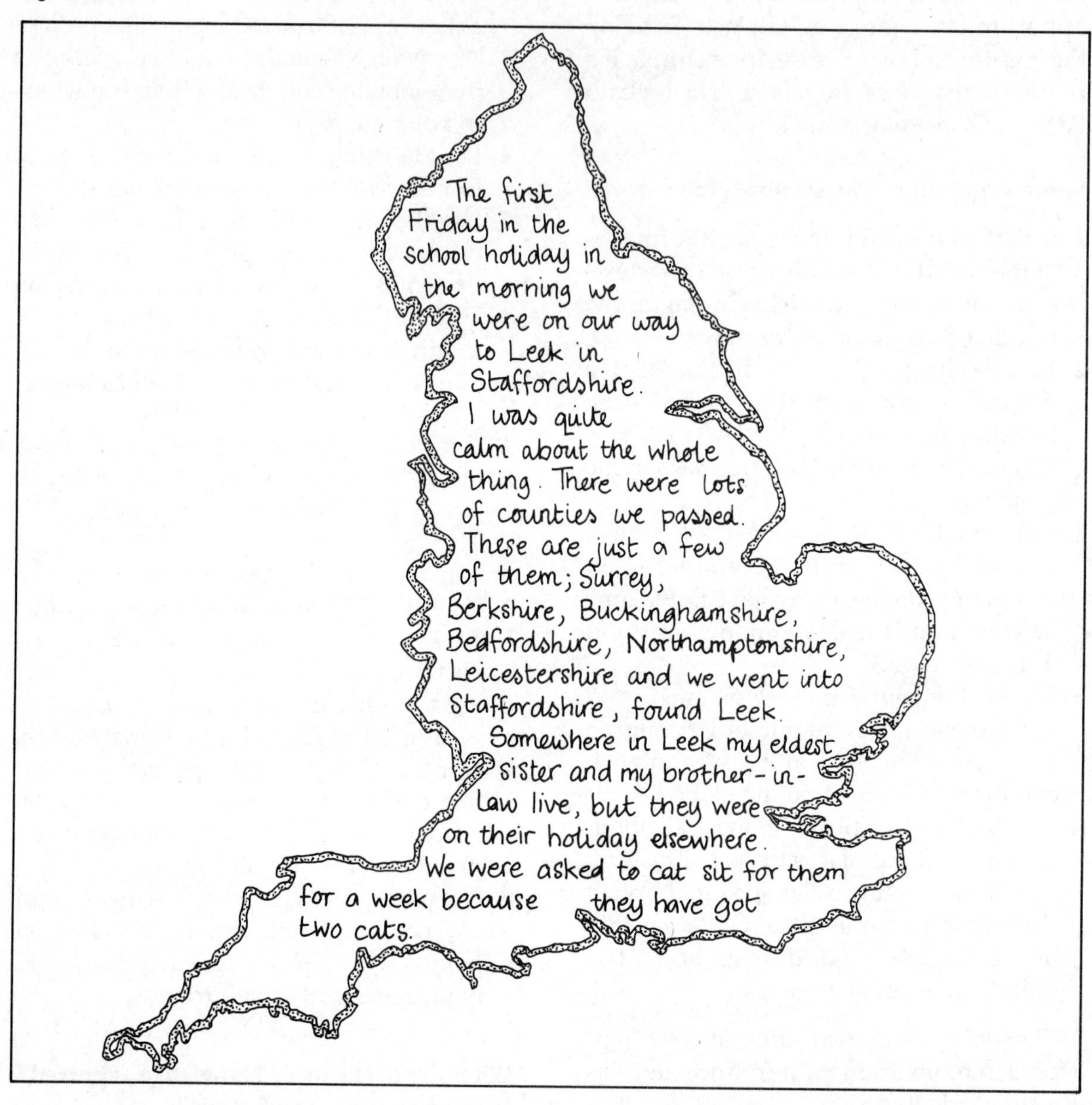

talked about 'sending the maps home for homework'.

Little understanding of the relationships between locations and spaces results from such learning. Take the real example of an eleven-year-old pupil who in a test located various places on a blank world map in the correct position relative to the edges of the paper, but with the world map upside down! Children have to build up an understanding of the relationship between world and country scale, space, and places within the space to grasp atlas work.

Although there may be a case for the occasional 'fill in these places on the map' assessment with older children, places and countries should always be referred to in context. That context does not always have to be in specific geography teaching. Every time a locality is referred to in geographical learning, it should be identified in the atlas, on a wall map or globe, as appropriate.

Upper juniors should have an atlas each or, at least, one between two. This ideal may well take time to achieve due to annual budgets and the need for a cycle of replacing worn stocks.

Atlases should be easily accessible to children as independent learners. Avoid locking them up or storing them in a central resource area. Avoid sharing atlases between classes. Children need to keep them in their trays, drawers or desks for quick, independent access. If the school budget has not yet permitted this critical provision, and central storage is necessary, there must be an atlas in the class library for quick reference in the interim.

A fun atlas quiz is a good occasional activity for the whole class to assess how children's knowledge is progressing.

If you are pointing out a place on a wall map or on an overhead transparency, or if a pupil is doing so, have the rest of the group locate it in their atlas, too.

- Who can do it first?
- Who can help their neighbour find the country shape and name?
- Who can show the shape and name to their neighbour?

Ask children to work in pairs or small groups, taking it in turns to quiz the others.

- On which page is the map of Europe?
- On which page is there a map of Australia?
- On which page is *just* Australia shown?
- On which map is the mouth of the Amazon shown most clearly?
- Find London on the map on page 20 and point it out to me.

Teach atlas skills; they do not get absorbed incidentally. The use of the contents list at the front and the index at the back has to be taught, in the same way that information reference book skills have to be taught. The learning and reinforcement of these study skills taught in this geographical context can develop the English curriculum and can be recorded as such. Practising these skills can be made fun for children by turning it into a whole-class speed game for ten minutes or so at periodic intervals during a term.

- Who can be first to tell me the page on which I can find a map of Europe?
- Who can find a map of England which shows how high the land is, using the contents page to start with?

Latitude and longitude also could be mentioned for able pupils in upper key stage 2, although technically specified in the key stage 3 PoS. Many atlases explain how the latitude/longitude grid is imposed on the globe and how to use the system, or familiarise yourself with it first. Latitude and longitude are in fact much more complex than OS map-type grid referencing. It is useful to make the link between the two systems, but tell the children that latitude and longitude always start from the Equator and use north/south references first, before east or west of Greenwich. 'Along the corridor and up the stairs' or two cardinal directions is no use for latitude and longitude!

Remember that in both globe and atlas work two different concepts are being developed – one relating to space or area, the other to specific locations or place. Land masses, sea and countries occupy SPACE. Settlements appear first as points on maps and therefore as points located – PLACE. Some linear features can also be recognised depending on sale – chiefly the larger rivers such as the Amazon or the Nile. Roads and railways may, of course, be shown on country scale maps.

As teachers we need to be aware that these spatial and locational concepts are difficult for children to grasp. Often well-planned lessons can go wrong because we have assumed that younger juniors understand map vocabulary and its concept when most of them do not. Many children may need to be told that the 'dot' or 'point' on the atlas map stands for a type of settlement. They may need to know that the lines joining them are types of roads or railways, and which words refer to the name of the town or village, the river or the hills. A whole host of coding is written into atlas maps which as adults we take for granted, but which for children may be incomprehensible, even with the key.

Figure 8.18 summarises the development of map skills against levels as a check for your planning.

Figure 8.18

Revised National Curriculum Geography: The development of map skills including all NC requirements

	Location	Representation	Distance	Perspective	Style	Drawing	Map use	Map knowledge
Infants (R and Y1)	Follow directions: e.g. up and down, left and right, behind, in front of	Use own symbols on imaginary maps	Use relative vocabulary bigger/smaller like/unlike, etc.	Model layouts; draw round objects to make a plan	Extract information and add to picture maps; use globes	Draw picture maps of imaginary places and from stories	Talk about own picture maps	Constant reference to countries, seas and place names on the globe and large maps
Infants (Level 1/2)	Follow directions: north, south, east and west	Use class agreed symbols on simple maps	Spatial matching; begin to match the same area, e.g. continent, country, on a larger or smaller scale map	Look down on objects to make a plan; plan view needs teaching	Land/sea on globes; teacher-drawn base maps and large-scale OS maps	Make a representation of a real or imaginary place	Follow a route; use a plan; use an infant atlas	Locate and name on map of UK major features, e.g. seas, rivers, cities, home location
Lower junior (Level 3)	Use letter/number coordinates and four compass points	Introduce need for a key and standard symbols	Spatial matching, boundary matching – identify the same boundary, e.g. country boundary on a different scale map	Draw sketch map from high view-point; add slope and height	Identify features on oblique aerial photographs	Make a map of a short route with features in the correct order; simple scale drawing	Use larger scale map outside; use maps of other localities	Progress towards ...
Upper junior (Level 4)	Use four-figure coordinates to locate features on a map; use eight compass points	Draw a sketch map using symbols and a key; awareness of some OS symbols	Measure straight line distance on a plan	Increasing use of plan view mapping	Use index and contents page in atlases; medium-scale OS maps	Draw a variety of thematic maps, based on own data	Compare large-scale map and vertical air photo; select maps for a purpose	Identify points of reference specified on Maps A (UK) B (Europe) and C (World)
Access to key stage 3 programme of study for level 4 and beyond	Use six-figure grid references to locate features on OS map; latitude and longitude on atlas maps	Use OS standard symbols; develop use of atlas symbols	Scale reading and drawing; comparison of map scale	Use models to introduce idea of contours; submerge or slice; interpret relief maps; identify relief features	Interpret distribution maps; concept of globe as flat map	Draw scale plans of increasing complexity	Follow route on small-scale OS map and describe features seen	Begin work on identifying points on Maps D (UK), E (Europe) and F (World)

Originally based on the matrix in Mills, D. (ed.), *Geographical Work in Primary and Middle Schools* (Geographical Association, 1988)

9

GEOGRAPHICAL FIELDWORK

A philosophy of fieldwork

Fieldwork means active geographical learning outside the classroom. It has long been acknowledged by HMI and teachers with enthusiasm for primary geography as being a basic tool to foster sound learning and understanding. Fieldwork, like assessment, is not a bolt-on accessory; it must be an integral part of planning for National Curriculum geography. There are many opportunities which can be taken within any key stage plan to develop fieldwork skills. Well-planned and executed fieldwork enhances the children's learning. Some key questions and focus questions used in units of work will require fieldwork.

Our primary school pupils are entitled to continuity and progression in fieldwork for these reasons:

- Good practice in geographical topics requires it
- It helps develop process learning
- It is active learning
- It is motivating
- It is enjoyable
- It can be fun
- It can be exciting
- It helps develop a feeling for the environment
- It aids children's personal and social development
- It can aid cross-curricular learning
- Occasionally it can involve a small physical challenge

What other area of the curriculum lends itself so easily to first-hand experience as geography? Geography is all around us *now* – a wonderful opportunity to be used as a teaching resource! Perhaps it is just because the landscape and the processes, natural and human, which operate upon it to produce patterns are always there, that both children and adults can take it for granted and ignore geography. Fieldwork is an essential skill to develop a child's sense of place and space, helping to answer the basic question 'Where am I?'

In the current climate, it is actually easier to choose not to take children beyond the confines of the classroom for active learning. Legal, financial and organisational considerations can overwhelm even the keenest teacher, so we need to be sure that our rationale for fieldwork is a strong one. We never need to convince our pupils that fieldwork is a good practice – they love it! But we may need to convince others that spending time *beyond* the school grounds is essential to good practice in primary geography, history and science. Head teachers, governors and parents will need to be convinced of the value of fieldwork, because of cost and safety issues.

Head teachers will have further concerns relating to supply cover costs to increase staff/pupil ratios, their own possible time involvement to lower these ratios, and general use of children's time through the whole curriculum. Other colleagues who have not necessarily been involved in fieldwork to date will need convincing, because they may feel it will involve yet more work on their part.

However, the geography Statutory Order does support good practice in fieldwork, because it would be impossible to deliver the programme of study and assess children's work without engaging in first-hand, enquiry-based fieldwork in, at the very least, the school grounds and the local area.

The programmes of study for key stages 1 and 2 state quite clearly in paragraph 3b that pupils should be taught to 'undertake fieldwork'.

The programmes of study are legally binding, so fieldwork is statutory. However, there is no stated time requirement for the fieldwork element of primary geography in the Statutory Order, as this would also be legally binding and put schools in an impossible position. But the National Curriculum *Geography for Ages 5–16 Final Report*, June 1990, suggested this as a minimum entitlement for work further afield:

> "*Every child in years 1 to 6 should experience at least one day visit per year to a location beyond walking distance from the school, so that the contrasting geography of another locality can be explored.*"

As a guide, the authors suggest that at least one fieldwork visit beyond the school gate every term is a realistic target. It is taken for granted that the duration of such fieldwork will vary according to the children's age and needs, and will include the demands of other subject fieldwork.

Both good practice in primary geography and the geography Statutory Order, reinforce the fact that fieldwork must be integrated into the geography curriculum. Like mapwork, it should not be done as a 'parcel' of learning, but should be a skill practised in the context of a place, and one or more of the physical, human and environmental geography themes for key stages 1 and 2.

The integration of fieldwork skills means that they must be carefully planned for, carried out and followed up back in the classroom. Figure 8.1 indicated the minimum of fieldwork skills and equipment that need to be planned for in the primary school.

Continuity and progression in fieldwork

The key to achieving continuity and progression in fieldwork is setting learning objectives through key questions and the following criteria. Progression in fieldwork is achieved over key stages 1 and 2 by building up conceptual development through work in the following areas.

1 The precision and detail in collecting data
2 Depth in the amount of follow-up analysis of data.
3 The scale of locality in which pupils do the fieldwork, moving from the familiar to the unfamiliar.
4 The complexity of ideas and techniques pupils use in fieldwork.

Here are some examples to clarify what is meant by this:

1 Increasing the precision and detail required in collecting, handling and recording data could involve:

- A key stage 1 pupil doing a building

survey in the local area being required to observe front door numbers and colours and to ask the adult accompanying the group to record these on a large-scale map
- A key stage 2 pupil doing a similar survey being required to do their own recording and accompanying it with several sketches of complete houses, labelling all the house types (terraced, detached, etc.) and house features such as slate tiles, brick, weatherboarding, double-glazing.

2 Increasing depth in the amount of follow-up and analysis of data could be developed in, for example, weather work, which lends itself to such progression. Development can be observed in:

- Infants noting and displaying their weather observations in a class bar chart and writing a sentence to assess whether there are more rainy days than sunny days in one week
- Juniors inputting their observations made from accurate measurements into a computer database and examining the resulting computer graphs to see if a pattern can be observed – the resulting pattern could be compared with the one from the previous year if statistics were kept for the same period.

3 Increasing the scale of the locality in which children do the fieldwork means moving from the familiar to the unfamiliar, involving:

- For the youngest children, fieldwork activities in the context of the school building and site – many studies can be made on a small scale in this safe and easily accessible location
- For children progressing through key stage 1, fieldwork in the locality around the school
- For children progressing through key stage 2, fieldwork in unfamiliar places away from the school, elsewhere in their home area, or much further away in the UK – some primary schools, particularly those located near the Channel ports, may even take a day trip to continental Europe or take a residential trip to a twin town or contrasting locality.

The length of time a field visit needs to take should, in theory, relate to the purpose of the fieldwork. Where costed transport is necessary, practice sometimes dictates that some geography-focused fieldwork needs to be included alongside some other activity such as historical or scientific fieldwork. In the past, the idea of fieldwork in the primary school developed from the post-war school outing to a well-known location a long way from the school, and we have inherited the feeling that a field trip needs to last for a school day.

The fieldwork activity needs to last as long as the teacher and pupils plan for it, although economics may distort best practice. It could be as short as one hour investigating their local shop. Progression is achieved by increasing the length of fieldwork experiences and moving to unknown, more distant, locations.

Figure 9.1 shows diagrammatically the progression and continuity of fieldwork experience which good practice in primary geography requires. The Statutory Order does not mention residential fieldwork because the charging issues follow the Education Reform Act 1986, would make it impossible to comply with them. Nevertheless, the requirement to study a contrasting locality in the UK will only be motivating for most children if they can visit such a distant place and study it at first hand.

4 Increasing the complexity of ideas and

Figure 9.1

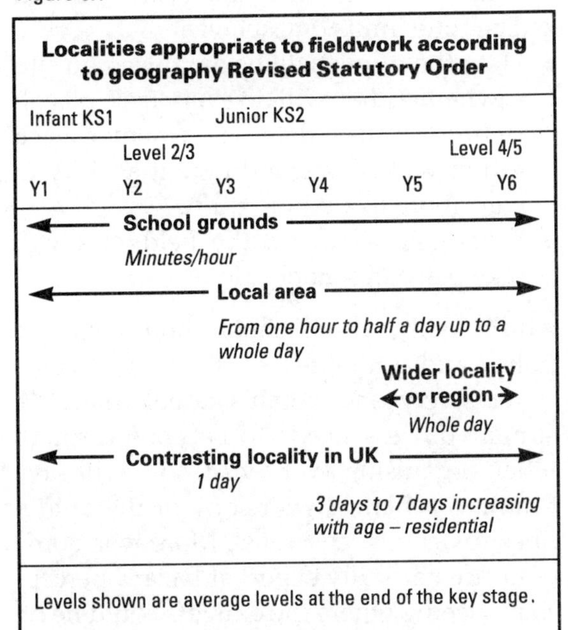

techniques can again be illustrated by weather work. The Statutory Order requires that infants 'should be taught about the effects of weather on people and their surroundings'.

- Infants begin to notice the effects of rainwater on different surfaces such as soil and tarmac, and observe and experience daily weather patterns – both can be done in the school grounds
- Juniors through KS2 PoS 8a study microclimate. Microclimate is a new concept for many teachers as well as pupils. It means studying the effect of site on weather measurements. Different temperature readings will be taken at different mini-sites in the school grounds. Those taken at the same time on a sunny site and in the shade of a building will vary. Those taken on a windy day in an exposed playground site in the sun will be lower than those recorded in a sheltered sunny site. Soil temperature readings will vary when taken under bushes and in a flowerbed. Such different readings will need to be collected and logged and their locations mapped as part of a wider investigation. The most sophisticated level of microclimate work for schools lucky enough to afford the technology (several hundred pounds in cost) is to use computer remote control sensors to log temperature, windspeed, etc. in the various mini-sites, but this progression should only be undertaken by pupils who have used and understood standard equipment for themselves first. This could be a sharp-focus enquiry on where to site some new playground seats or as part of a science/geography-focus weather topic.

Continuity in fieldwork goes hand in hand with progression. It is achieved by practising the same skills but at ever-increasing depth. The transference of skills learnt in one situation to another location is very important. The use of the compass in a fieldwork location, even though it may be used every time a class goes into the local area or school grounds, needs to be practised in a new and distant location. The child is often excited in an unfamiliar location and will need the confidence to repeat an activity already carried out in a secure environment by repeating it, and relating to it in a much wider context. Estimating north, checking this with the compass and pointing out landmarks and their direction relating to north, need to be applied at the top of that exciting castle, too, so that the children release that the skill is transferable.

Continuity of location is also important. Fieldwork in the school grounds and local area will be ongoing throughout key stages 1 and 2, although new areas will be brought in with progress up through the junior school, as in Figure 9.1.

Greater depth and breadth of study in any continuous use of the same area is essential, reflecting the importance of whole-school planning and the passing on of records. A local stream examined as a geographical feature – its banks, meanders and direction of flow – in the lower junior years may be used with the upper juniors in the context of a pollution project to investigate pollution and conservation issues as part of a wider and local area study. KS2 PoS 10a, 10b, 7b and 5a could all be touched upon.

Some examples of the kind of fieldwork activities which can be carried out in the local area in particular environments are given in Chapter 10.

Safety and organisational issues

Well-planned and effectively organised fieldwork will be one of the most satisfying professional experiences a teacher can have. It will prove extremely motivating and enjoyable for pupils by providing learning achievements and will win the admiration and respect of accompanying parents. On the other hand, fieldwork undertaken lightly will be a nightmare for all concerned.

Safety concerns and organisational issues are interlinked. For example, safety factors impinge on the organisation of pupils working in groups, which must be planned well in advance. The two areas will be dealt with together here, and they apply to all fieldwork whether it lasts an hour in the local area or a week in a contrasting locality.

Fieldwork will be a crucial part of the topic or theme the class is studying. It will most likely be part of an enquiry, so we shall need to plan:

- What key questions the pupils will be asking
- What tasks they will be doing on site
- What is to be done in the classroom before the visit
- Whether there will be one visit or several
- The time limits on each visit
- How the pupils will be organised on site – whether they will do every task, all collecting similar data, or whether each group will pursue a different activity to contribute to a whole study
- What follow up on the fieldwork will take place in school.

Additionally, your school should have a policy and guidelines on fieldwork which cover safety and organisational issues. A similar process needs to be gone through when organising every visit, so it will save time if teachers have a copy of the guidelines to act as a check-list. Moreover, some of us are naturally expert at hazard perception – seeing where the dangers could lie on a visit – whereas others amongst us may not have had a great deal of experience. It is hoped that the following may prove helpful for primary geographical fieldwork.

Guidelines for primary geography fieldwork

The essential formalities

1 Think through the aims, objectives and broad learning outcomes of the fieldwork.
2 Check that you are familiar with any school or LEA documents regarding pupil–teacher ratios, costing, insurance, etc.
3 Ensure that you understand charging legalities and your own school's charging policy; obtain costings.
4 Seek permission from the head teacher and governors to take pupils off site for fieldwork.
5 Arrange dates and times.
6 Consult any colleagues who need to know.

7 Arrange for extra adult help.
8 Obtain parental consent on a standardised form. *These forms must be carried off-site by the party leader on any occasion, whether local or residential.* Many schools have two types of consent form: those which are filled in once a year to give blanket consent for local visits over the year; and those which are filled in specifically for residential visits or those of longer duration. The books mentioned at the end of this fieldwork section provides examples of such forms. Remember that consent forms for residential trips need to include any notes on:

- Consent for surgery if action is considered to be urgently needed
- Religious limitations on surgery
- Details of contact numbers of parents or guardians.

Prepare yourself

1 For your own peace of mind, make yourself aware of pupils' up-to-date medical and dietary details. This is essential for residential work, but good practice also for day fieldwork visits. Bee-sting allergies can be as dangerous 15 minutes away from school as 100 miles, as can diabetes and asthma! Religious beliefs may forbid pupils to eat certain meats such as beef or pork, or any food in which beef and pork products are ingredients.
2 Make a visit to the fieldwork site. *This is essential preparation.* For residential fieldwork at least the party leader should have made a visit in advance
3 Collect any relevant materials on your pre-visit. List and assess potential hazards and organisational details.

- Where can the coach drop off and pick up safely?
- Do you know where the nearest toilets for pupils are?
- Which site will be suitable for lunch?
- Is there a telephone box nearby in case help is suddenly needed?
- Are there any points which will present difficulties for special needs pupils?
- What alternatives or modifications can you make to your plan if the weather is bad?
- What equipment might you need to bring?

4 Sort out costing and insurance procedures with the head teacher.

Prepare your helpers

1 Take the time to explain the purpose and expected outcomes to other colleagues or, more likely, parent helpers. *They cannot be effective if they are not fully aware and involved. Work outside the school site is not a holiday for helpers; make them aware of this!*
2 Ensure that they know what to do in the event of an accident and who has final responsibility for decisions on the trip.
3 Ensure that they know what your discipline standards are and what their role is regarding them.

Prepare the parents

1 Inform the parents in writing of the field trip location, date, time and nature of cost, voluntary or compulsory, and method of optional payment.
2 For residential trips, arrange a parents' meeting to explain supervision arrangements, clothing and footwear needs, spending money arrangements and expected behaviour standards. Allow parents to air any concerns. Face-to-face communication is usually far more beneficial for everyone, but provide a written checklist for parents at the meeting, too, and see that any absent parents receive it.

Figure 9.2

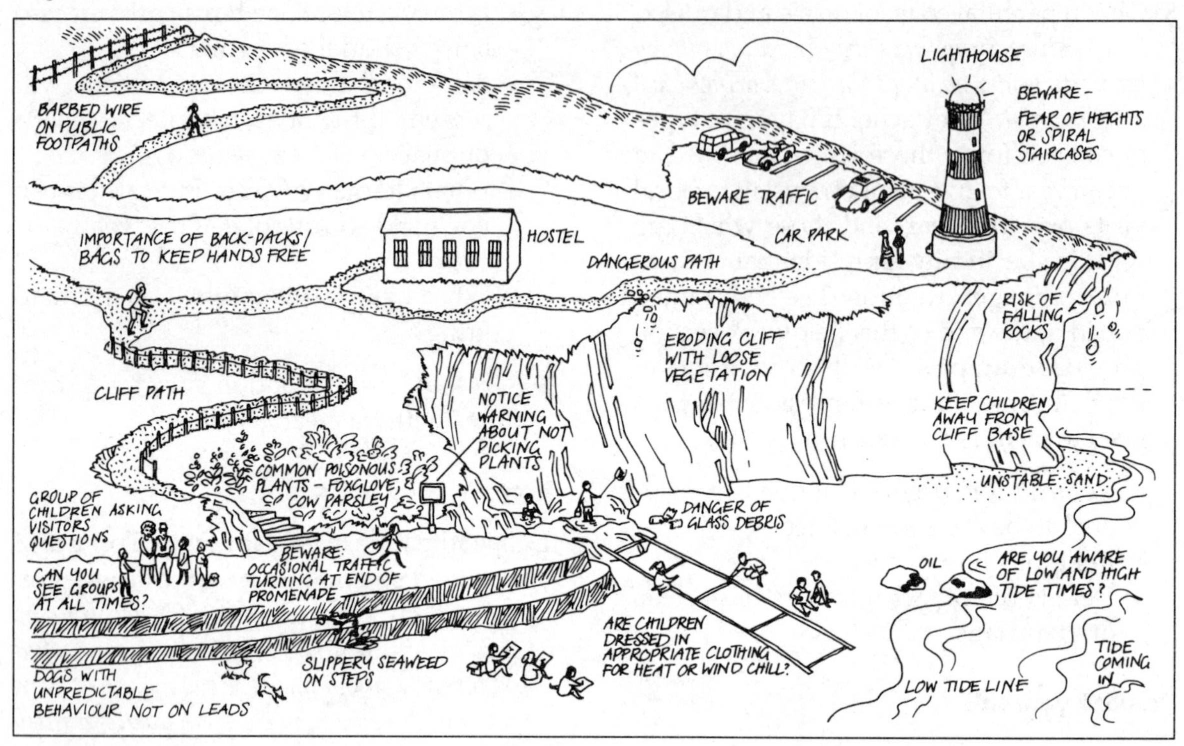

Figure 9.3

Prepare your pupils

1 Carry out the work you have planned to do in the classroom before you go.
2 Practise any necessary fieldwork skills in the school grounds before you go according to your knowledge of the children's experience.
3 Ensure that the children understand why they are going to do fieldwork.
4 Ensure that children know what their tasks are to be on the visit. Make them aware that you have prepared parent helpers. If any child's behaviour threatens the safety of the party, you are entitled not to take them off site. You must enlist your head teacher's support. Due discussion with parents in advance about expected standards of behaviour is reasonable. If a head teacher is adamant that a pupil should go, then the teacher/pupil ratio must be raised to your 'safe' level.
5 Go through your code of behaviour with the pupils.
6 If travelling by coach, anticipate travel sickness arrangements. Every child should take a spare plastic bag. The coach should also be provided with a bucket, toilet rolls, paper tissues, paper towel roll and air fresheners.
7 Check up on other medical arrangements – inhalers for asthmatics, etc.
8 Ensure that pupils are aware of any hazards, and that they know who to turn to in the event of a problem.
9 Remind pupils to bring suitable clothing, food and equipment.

Managing the fieldwork successfully on the day or residential journey

1 Be aware of the weather forecast.
2 Check names and numbers of pupils; see that all staff have a list of names.
3 Check that everyone is aware of the 'chain of command' – who has final responsibility during fieldwork.
4 Check that you have appropriate emergency contact numbers: for a day visit, school and head teacher at home; for a residential visit, the above, plus one of these – local adviser, inspector or director of education.
5 If you or the school possess a mobile phone, ensure you take it with you! Surely this use of a mobile phone for emergency communication is crucial for peace of mind nowadays.
6 Check that pupils needing travel sickness pills have taken them and given all medication to you. Allocate seats at the front of the coach. Check that everyone has a spare plastic bag.
7 Check that you have remembered the whistle and the first aid kit; a small pocket kit is necessary on a local walking visit.
8 Have a cash float and/or cheques for entrance fees and emergencies.
9 Remind pupils of hazards and code of behaviour.
10 Check that pupils have equipment as organised.
11 Upon arrival, assess the site. Things may have changed since your pre-visit. Take contingency action if necessary.
12 *Have a safe, brilliantly organised and enjoyable field visit!*

And now you have successfully returned ...

1 Carry out planned follow-up work.
2 Write, or see that children write, any public relations thank you letters.
3 Evaluate with the children what you all learned from the visit.
4 Keep any useful materials, especially those which will cut down on work for the next time.
5 Communicate the result of your work to

colleagues, other classes, parents and governors, as appropriate, to inform, encourage and celebrate successful work. Use displays in a central place in school, assemblies, parents' meetings, staff and governors' meetings, if appropriate.

Further details and advice about geographical fieldwork as it applies to the primary school can be found in the Geographical Association's *Fieldwork in Action* series no. 3 (see Chapter 12) and *Safety in Outdoor Activity Centres: Guidance Circular 22/94*, a copy of which was sent to all schools in September 1994 by the DFE.

Fieldwork skills and techniques

There is a strong link between mapwork skills and fieldwork skills. Many of the elements of mapwork skills needed for progression in mapwork have been dealt with in Chapter 8, for example direction and compass use. Fieldwork skills and techniques are concerned with the observation, use of equipment and recording data through the enquiry process. When guided by the right key questions, observation and recording will develop a sense of location, space and place and build up a knowledge of patterns which can be recognised in the physical and the human environment. Observation and recording will also lead pupils to recognise the processes which are taking place around them, thereby developing their understanding of patterns and processes through first-hand experiences of similarity and difference.

Progression in landscape sketching

In asking the key questions – How can I record what I see here? What is this building like? What can I see from this viewpoint? How is the land used in this place? – pupils will find the skill of landscape sketching helpful to record information.

Technically speaking, drawing a fieldwork sketch at its highest level – for example, 'an annotated sketch ... to record and interpret a landscape' – is not a statutory requirement until key stage 3 of the geography Revised Statutory Order. This skill has a progression within it which starts at key stage 1 with an activity which many infant teachers will recognise – the sketching and labelling of a house. Many geographically- or artistically-minded teachers have always encouraged landscape sketching, so many primary school pupils transfer at age 11 with this geographical skill highly developed. A geographical landscape sketch is *not* a work of art. It is a sketched drawing of a view which – and here it differs from the artist's aim – will communicate to whoever looks at it the main features and use of the land sketched. Geographical vocabulary is used explicitly to label or annotate the sketch; this would be unlikely in a sketch which was purely artwork.

Landscape sketching is not an easy skill. Many simple steps can be practised in the local area on the way to landscape sketches which children can produce in contrasting environments on residential field trips. See Figure 9.4 and consider these points in relation to progression in landscape sketching:

- Increase the number and complexity of features in the views, but bring in 'outer limits' to the area to be sketched, as children will find that 'it won't all fit onto the paper'
- Children can make their own frame from card which has been subdivided so that they can concentrate on the view through each sector in turn
- Get children to focus on the horizons first; silhouettes first, detail last; details can always be drawn in school from a

photograph taken on-site or a commercial postcard, if one can be found showing the view from the same spot
- Make sure the feature and land use labels and positions are noted down – that won't be so easy to do back in school
- Encourage the oldest children to note down in which direction the sketch is orientated – use of the compass is reinforced yet again.

Differentiate by:

- Providing some pupils with a sketch showing the horizon and main features in silhouette. Ask them to fill in the missing features from their own observation or from a physical and human geography word bank provided
- Providing a partly completed sketch with gaps in the horizon and ask pupils to complete the sketch and label the missing features.

Pupils with severe drawing difficulties or spatial perception problems can work with a helper who scribes the words or by adding to the prepared sketch the observations they make.

Progression in land use mapping – the 'quadrat' technique

In asking any of the key questions – What is the ground used for? What are the physical features of the area? What grows here? How is the land used? Is the land use different here from the use elsewhere? – pupils will need to make a map as well as talk about their observations.

The term 'land use' in geography means describing and identifying what kind of vegetation, rock, soil or buildings covers an area. It leads towards recognition of whether the land is used for industry, leisure or residential purposes, and so on. So land use mapping always relates to physical or human geography or, more likely, to both. A land use map can be the size of an actual footprint or represent a large area such as a city or a county.

The programme of study for KS1 3d requires pupils to:

> "*Make maps and plans of real and imaginary places using pictures and symbols, e.g. a plan of their route from home to school.*"

Like landscape sketching, the skill of sketch mapping, or free-hand map making in the field, needs a lot of practice. Again, it is the immensity of the sense of space that pupils find so difficult to commit to paper. 'I've gone off the edge of the paper', is a pupil's often-heard cry for help with free-hand map making.

By defining for children the limits the sketch map covers, we can help them to gain confidence in sketch mapping. A progression of activities over key stages 1 and 2 in the school grounds mapping land use will help them towards coping with less well-limited tasks in the upper junior age range.

The kinds of 'quadrat' activities described on page 111 will work towards achieving the programme of study requirements, as well as the other mapping activities already described in Chapter 8.

The idea of the quadrat technique has been usefully adapted for primary school fieldwork. Usually found in secondary school biology or geography, a quadrat – a square-metre frame divided into four sectors – is traditionally used to find the number and type of different plant species grown in a small area. The square-metre frame is cast at random on the ground, and the number of plant species found counted or estimated. A series of quadrat examples can be taken elsewhere in the area at random, and the

Figure 9.4

Progression in landscape sketching

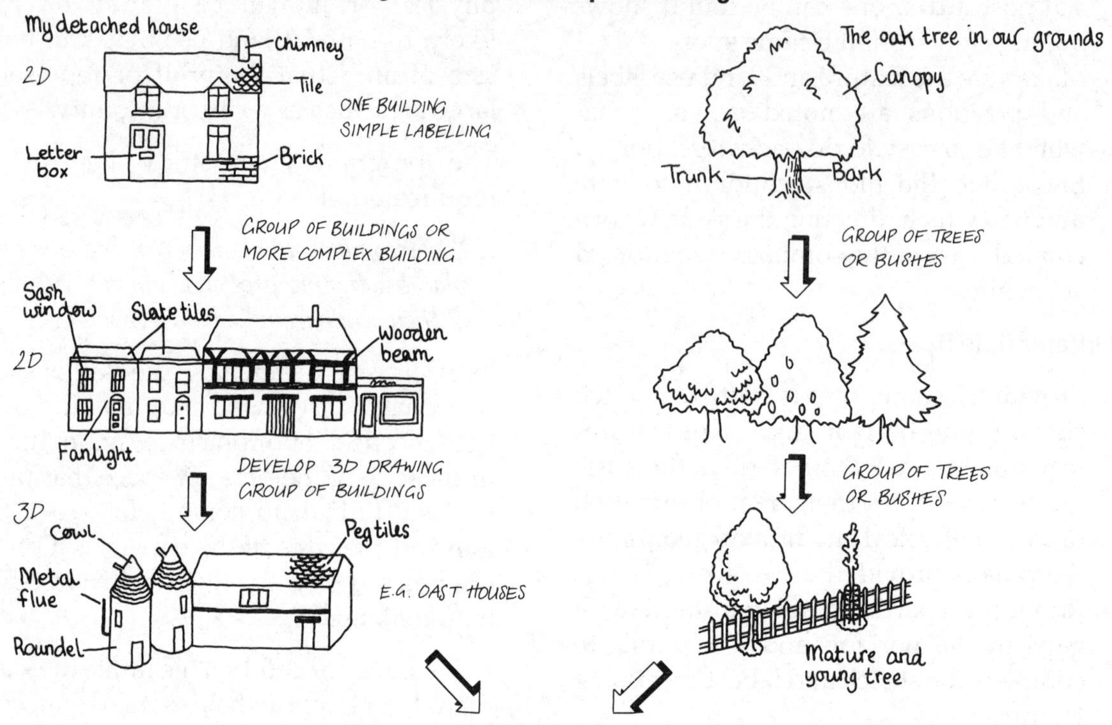

PARTLY COMPLETED SKETCHES

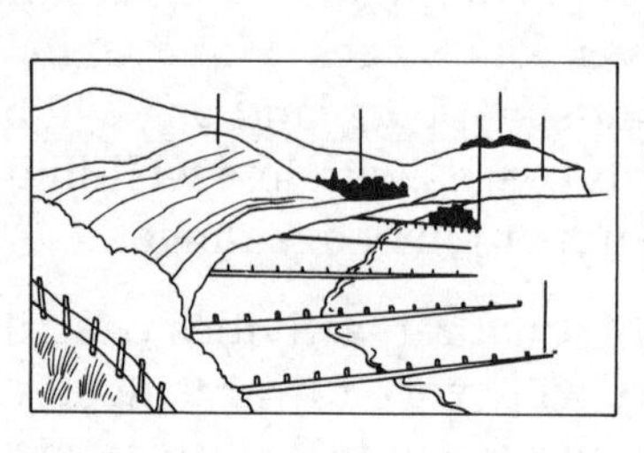

LABELS OMITTED. CHILDREN OBSERVE LANDSCAPE, IDENTIFY FEATURES AND LABEL ACCORDINGLY

LABELS PROVIDED BUT PARTS OF VIEW OMITTED. CHILDREN FOCUS ON SILHOUETTE OF FEATURES AND COMPLETE SKETCH

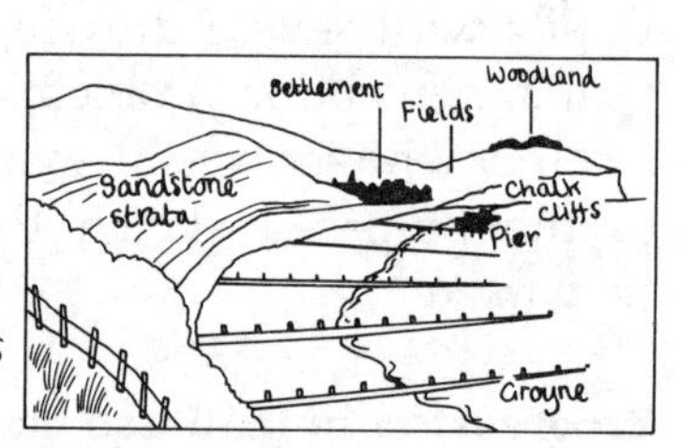

THIS KIND OF ACTIVITY CAN ALSO BE DONE STARTING WITH POSTCARD VIEWS AS PART OF THE PROGRESSION OR IF IT IS NOT POSSIBLE TO GET OUT TO A FIELD LOCATION

PUPIL SKETCHES AND LABELS WITH OR WITHOUT FRAME USE

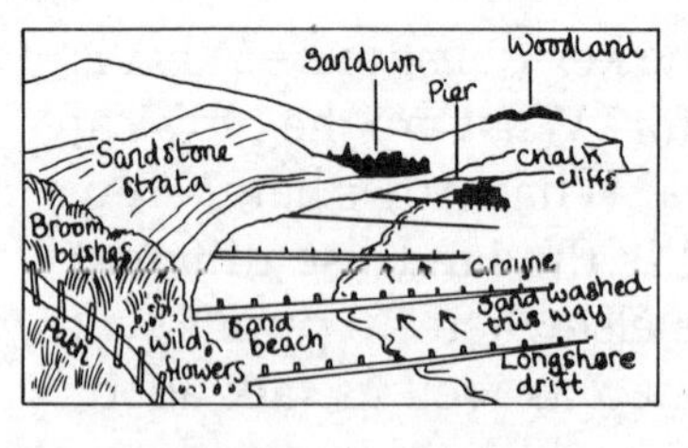

"FREE" SKETCHING

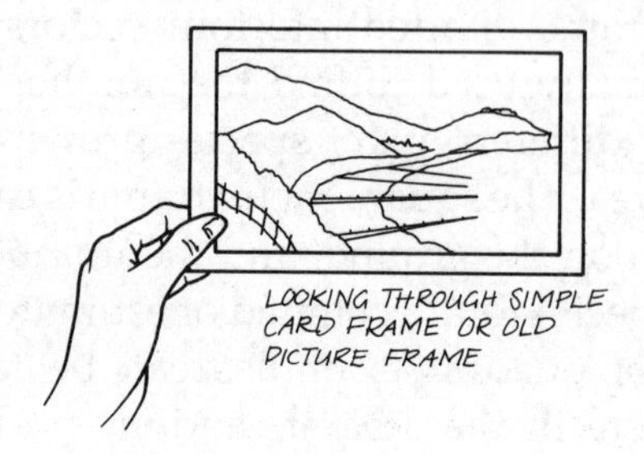

LOOKING THROUGH SIMPLE CARD FRAME OR OLD PICTURE FRAME

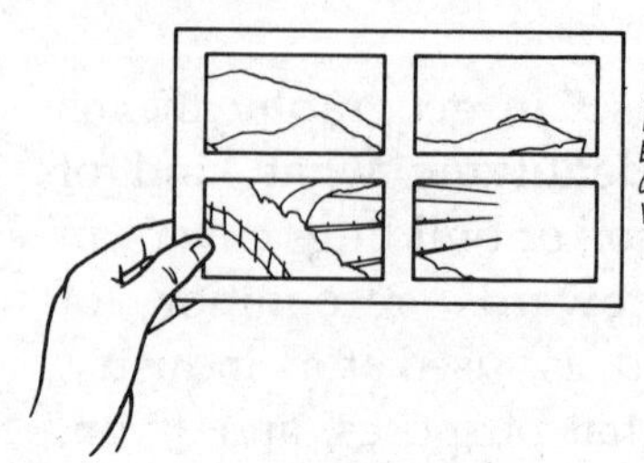

LOOKING THROUGH A DIVIDED FRAME ENABLES CONCENTRATION ON ONE SECTION OF THE VIEW AT A TIME

results averaged out to give a statistical result for the different type of plant cover.

For primary geography, it is not the number of plants which needs to be focused on, but the type and distribution of land use that pupils can observe. Using a square metre, with its connection with area work in maths, is for older pupils, but the progression map in Figure 9.5 shows how the technique can be developed at the primary school stage.

The school grounds provide the obvious environment for quadrat work, but it can be usefully done in the contrasting locality of a beach. Quadrat samples can be taken in different locations on the beach – along the high/low tide line, at the wave-cut platform, or on the wave-built terrace (see Glossary) – to build up a picture of the different rock/pebble, seaweed types and fauna distribution in the shore zones. Any good children's book on the seashore will provide teachers with the information needed to put them ahead of their pupils.

Hand in hand with quadrat activities, pupils should be able to develop the land use mapping of larger but still confined areas in the school grounds, focusing on observations, the representation of information, and the use of a key. The use of some quadrat activities should help children towards an easier, earlier grasp of land use recording and analysis.

A classroom extension of quadrat work for pupils who are mathematically able is to estimate land use from 1:25 000 or 1:50 000 maps. They can cover an area of the map with a plastic grid, shade different land use areas, and estimate residential coverage, and so on, in the local area.

Some further points about quadrats

1 How to make a quadrat – see Figure 9.6.
2 The inside area can be divided into more than four. Dividing it into a 10 by 10 cm grid is an obvious choice, with each pupil mapping a different mini-grid to recreate the whole square metre in the classroom.
3 Use quadrats to reinforce the micro-climate plant growth connection. Take quadrats under trees, in an exposed site, on a flower bed and so on, to draw out contrasts.
4 Change in land use over time can be mapped with quadrat techniques – over the seasons or over several years, or if the school grounds are changing due to wild area growth or additional playground space.
5 Quadrat work should normally be integrated into work about the locality of the school. A geographical topic or multi-subject-focused topic, drawing on science, environmental education and geography in the school grounds or the locality, are obvious ways of doing this.
6 Even an urban school with all-tarmac grounds can have a go at footstep and hoop quadrats – wet and dry tarmac or lichen- or moss-covered tarmac can be mapped. Try a vertical map – hold the shape against the wall.

Progression in height-measuring, slope and gradient work

The ability to estimate and measure height, and an understanding of slope and gradient are part of physical and environmental geography, particularly at key stage 2. Estimating the height of features and trying to obtain a reasonable measurement of height by scale drawing from fieldwork, enables pupils to record patterns in rural and urban environments.

The programme of study for key stage 2 requires pupils to use instruments to make measurements (PoS 3b). Clinometers are

Figure 9.5

Progression map for land use work

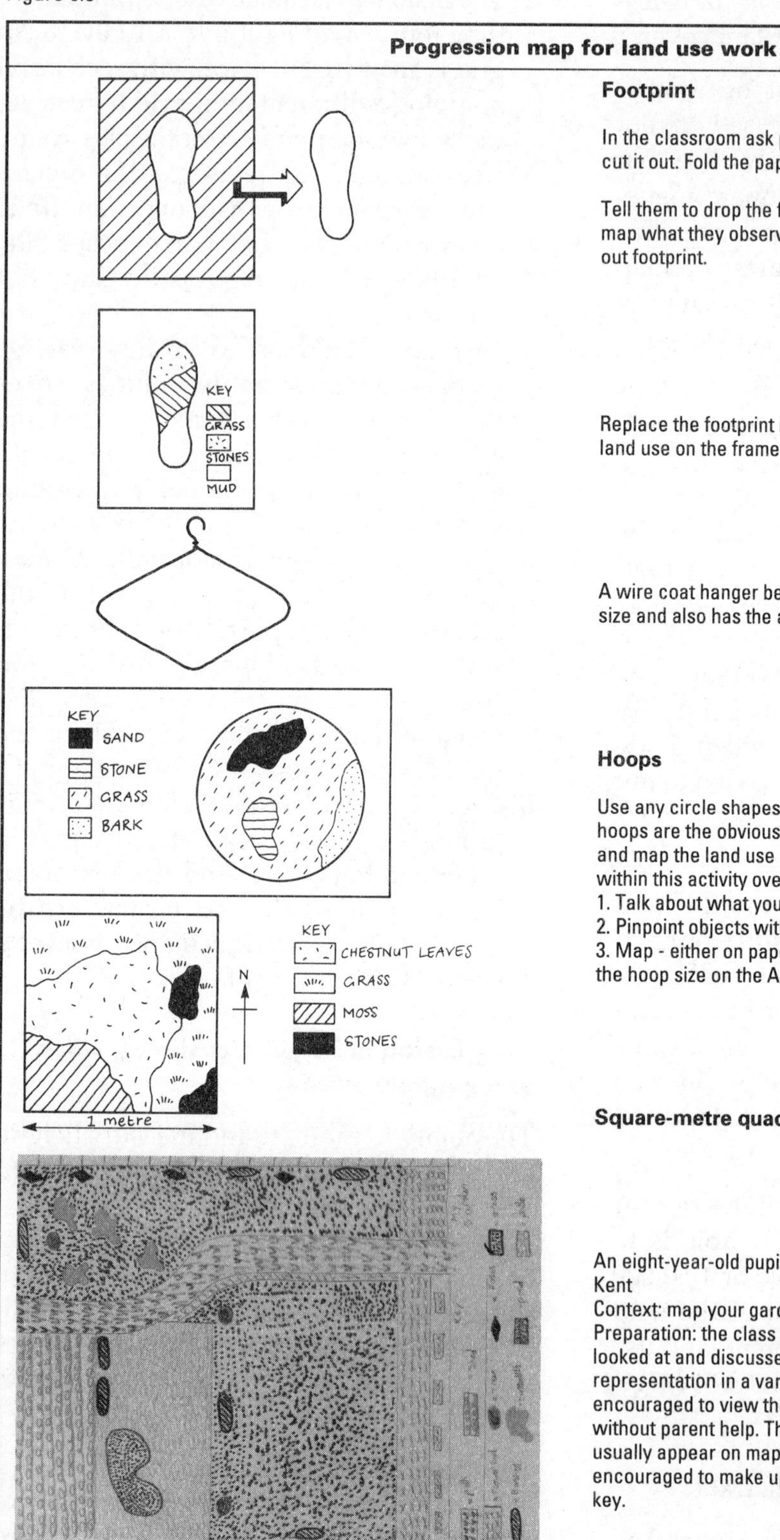

Footprint

In the classroom ask pupils to draw round their footprint and cut it out. Fold the paper in two to make the initial difficult cut.

Tell them to drop the footprint grid on the ground. They need to map what they observe in the space in the grid on to their cut-out footprint.

Replace the footprint in its grid frame. Record symbols for the land use on the frame.

A wire coat hanger bent out into a diamond shape is a useful size and also has the advantage of a handle.

Hoops

Use any circle shapes, all of which must be the same size. PE hoops are the obvious choice. Cast the hoop on the ground and map the land use inside on a circle of paper. Progression within this activity over time will be as follows:

1. Talk about what you see.
2. Pinpoint objects with markers without mapping.
3. Map - either on paper the same size as the hoop, or reduce the hoop size on the A4 paper.

Square-metre quadrats

An eight-year-old pupil's work from the Weald C.P. School, Kent
Context: map your garden.
Preparation: the class of top infants and eight-year-olds had looked at and discussed plan view and plan view representation in a variety of atlases and books. They were encouraged to view their garden from upstairs and work without parent help. The fact that moveable objects do not usually appear on maps was discussed. Pupils were encouraged to make up their own signs and symbols for their key.

one such instrument and may be used to compare angles of slope and to measure height where it would otherwise be impossible, for example with a cliff, tree or building, but this technique and its use is currently a mystery to many teachers.

Figure 9.6

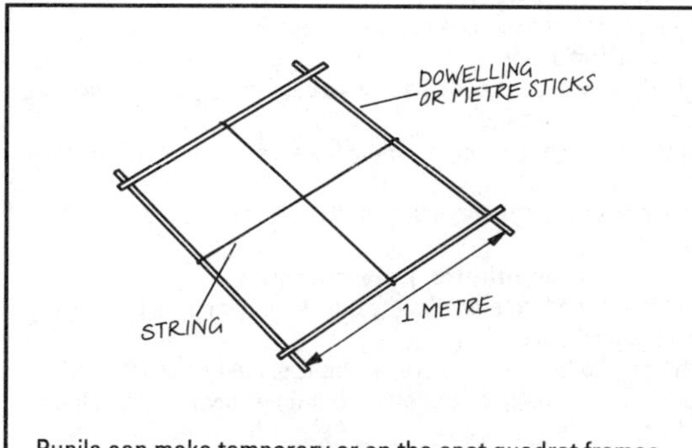

Pupils can make temporary or on the spot quadrat frames using four long sticks or skewers and four metres of string. More permanent quadrat frames can be made using dowelling and string. If dowelling is not available, four metre tapes can be laid on the ground with string to divide up the frame.

Lots of practice with a progression of simple activities built into key stage 1 and 2 work will lead to greater success in using and understanding the clinometer technique at the end of key stage 2. The progression map (Figure 9.7) is not intended to be exhaustive: you may have your own, similar tried and tested methods.

The context for this fieldwork height measurement technique is important. Sometimes it will need to be practised just as a skill to see that children can understand and use it in a different location, but often it can form part of an organised or incidental enquiry process. Here are some examples of ways in which it might be used:

- On a history field trip, a pupil asks the height of a castle wall while discussing attacking the castle's defences.
- If there are strong winds, would a particular tree fall on a particular part of the school, given a specific wind direction?
- If we planted a new tree in our school grounds, knowing the height it would grow to, would it dwarf the nearest tree?
- How high is the cliff on our local beach or in our field trip location? Can we make a scale transect – a cross-section – drawing of the cliff and beach as far as the low tideline?

Using a clinometer to measure steepness of slope

A clinometer can be used to used to find the angle of elevation of a slope. The larger the angles of elevation, the steeper the slope. Children can use the following method to obtain the angle of elevation for slope work.

Plant a rod at the top of the slope. Mark off a point on the pole at the same height from the round as the observer's eye. From the bottom of the slope sight this point through the sighting tube or line of the clinometer (see Figure 9.8).

On a school-made clinometer (see page 115), the angle made by the weighted string or thread with 90° while sighting is equal to the angle of elevation of the slope. (For commercial clinometers, follow the maker's instructions.)

Equipment for slope and height work

For younger children estimating slope, gradient and height, no costly equipment is needed. For accurate work with upper juniors clinometers are necessary.

A clinometer is basically a 180° protractor, used with a plumbline suspended from the centre of its straight edge. Commercial ones are now widely available in educational catalogues for about £10. However, it is possible for children to make them at virtually no cost, by one of the following methods:

Figure 9.7

A progression of height estimating and measuring activities

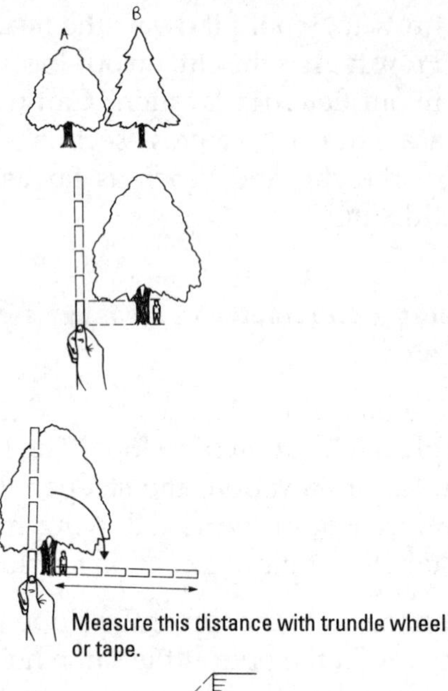

Measure this distance with trundle wheel or tape.

Suitable for infants

Direct comparison
Compare heights visually. Which building is taller? Which tree is taller? A or B?

Infants Lower juniors

Estimating height compared to a known object
1 Measure your partner's height.
2 Your partner stands against base of tree or building.
3 Hold a pencil at arm's length.
4 Site your partner with your pencil and hold your pencil so partner height fits with section of pencil length.
5 Estimate how many pencil lengths fit the height of the tree by counting pencil lengths up the tree.
The tree is six times my partner's height or 6 x 1 metre, 6 x 1.5 m.

Top infants Lower juniors

An extension of this method answers the question: If the tree fell, how much ground would it cover?
1 Pencil estimate the height of the tree as before, but this time in addition visualise these 'partner heights' paced horizontally outwards from the base of the tree.
2 Send partner to point on the ground which marks the end of the sighting having counted out and noted the 'partner heights' horizontally.
3 Measure the distance from the base of the tree to partner with a trundle wheel or tape.

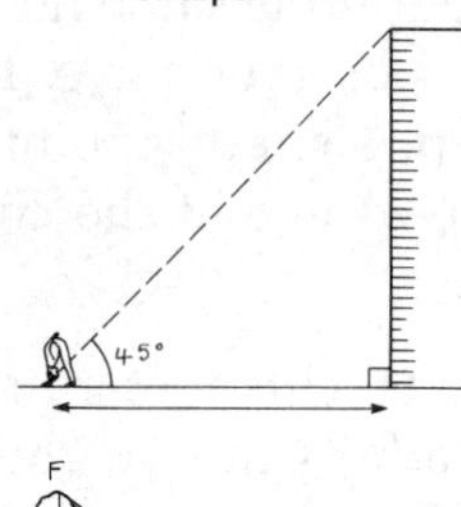

Using angles to find height

Walk away from the wall or tree until you can sight the top of it when you bend over and look back up at it through your legs. Mark where you are and measure the distance from there to the base of the building or tree. This distance is the same as the building or tree height because you are making a big isosceles triangle. You can use this method with or without the geometry for children if you wish.

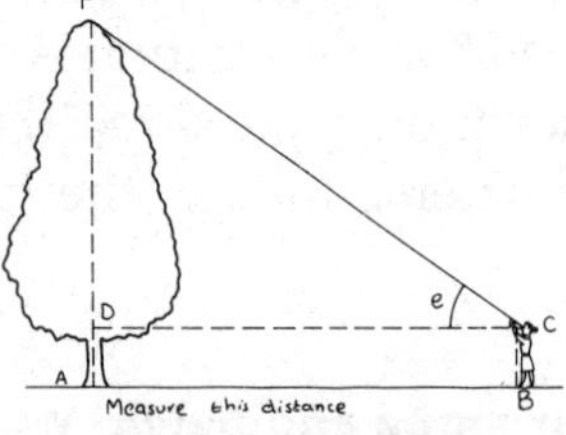

Measure this distance: *e* is the angle of elevation, the angle you look up through from the horizontal to site the top of cliff, building, tree, lamppost, etc.

Good upper juniors
Using field measurements, clinometer and scale drawing

On site

Children should work in pairs:
1 Estimate the height of the tree as a check for drawn results later.
2 Walk away from the base of the tree to a convenient point.
3 Measure distance A–B and note.
4 Measure height to eye level of pupil and record.
5 Sight the top of the tree, cliff, etc. with the clinometer and record the angle of elevation *(e)*.

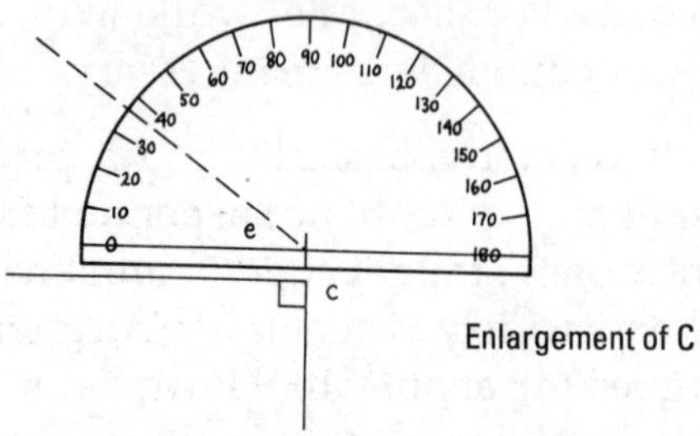

Enlargement of C

Back in the classroom

1 Choose a simple scale, e.g. 1cm=1 m.
2 On suitable graph paper plot distance AB.
3 Plot the eye height of your partner.
4 Complete drawing the rectangle ABCD.
5 With a protractor draw the 'e' – the angle of elevation from C, with the protractor base along CD. 'e' will usually be between 20° and 30°, depending on your distance away from the feature in the field.
6 Extend the line from C through the angle of elevation constructed in the last step until it hits a vertical line extended up through line A–D. This vertical line represents the height of the tree.
7 Now measure this line A–F. Transfer this length back to real measurement using your scale, e.g. AF = 8.5 cms so tree is 8.5 m. high.

Figure 9.8

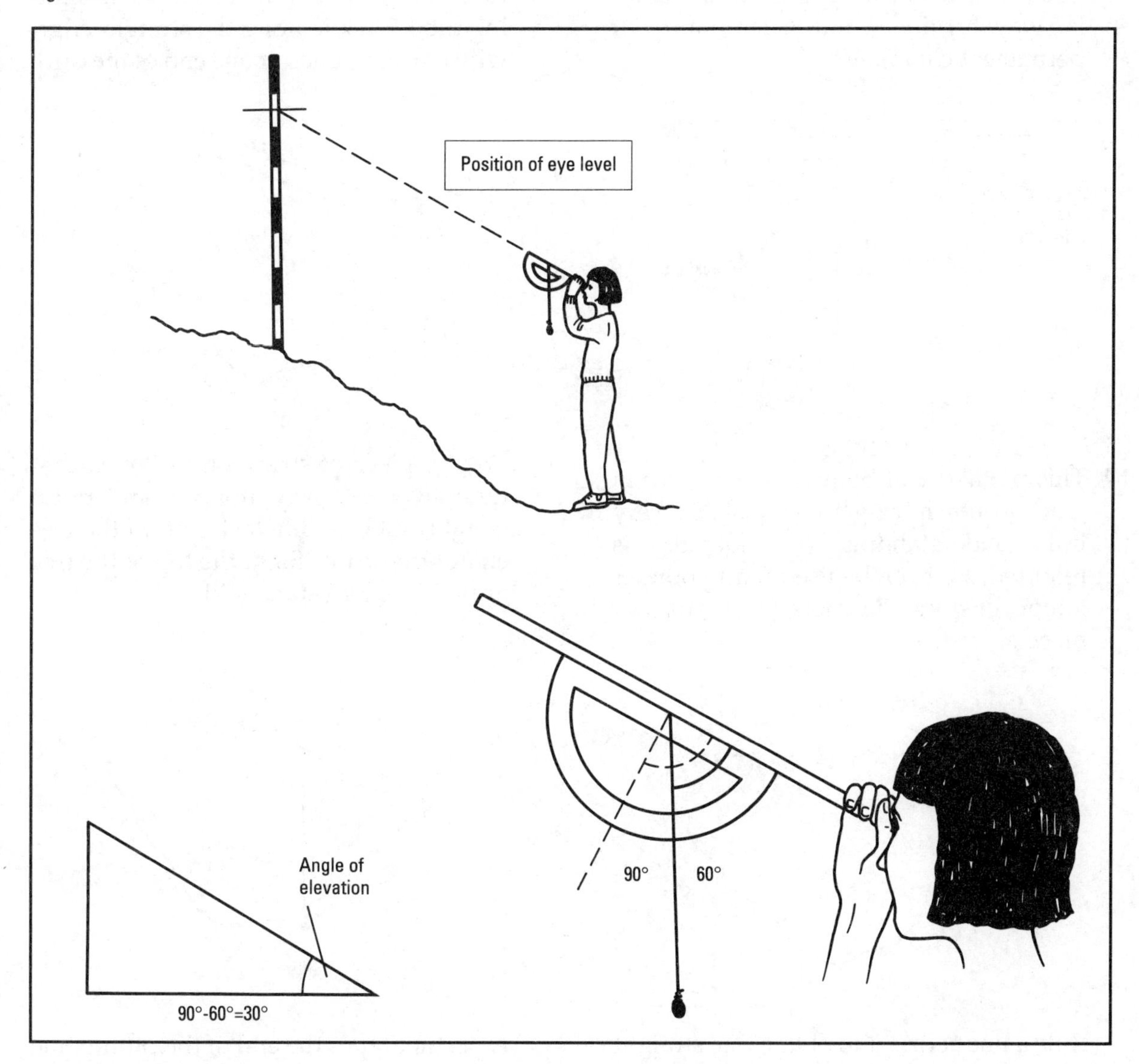

1 Photocopy a plastic protractor with white A4 paper behind it. Enlarge the image to the required size and cut it out. Stick it onto card, put a butterfly winged clip through the midpoint of the straight edge along the base line and attach a piece of string with a weight on the free end to form a plumb line.
2 Copy a protractor onto card, complete with angles marked in 5° divisions. Fix a plumb line as above, and stick a sighting, non-transparent straw along the straight edge base line.

3 Use your blackboard protractor in a similar way, turning it into a temporary or permanent clinometer.

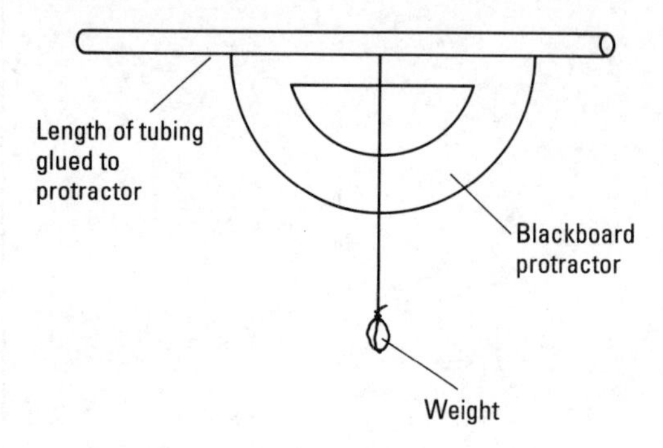

Roll up a sheet of paper to make a sighting tube. Stick it along the straight edge, with one of its ends at one end of the card.

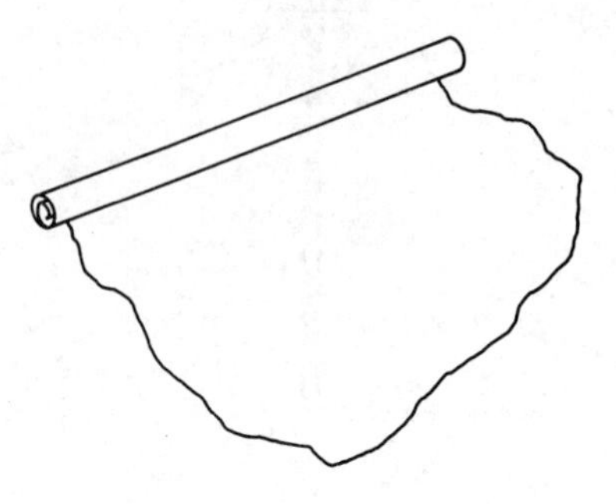

4 This method can be used with children who do not have geometry skills. Very little understanding of geometry is required, so it can be used with younger juniors or slower learners. Start with any piece of card.

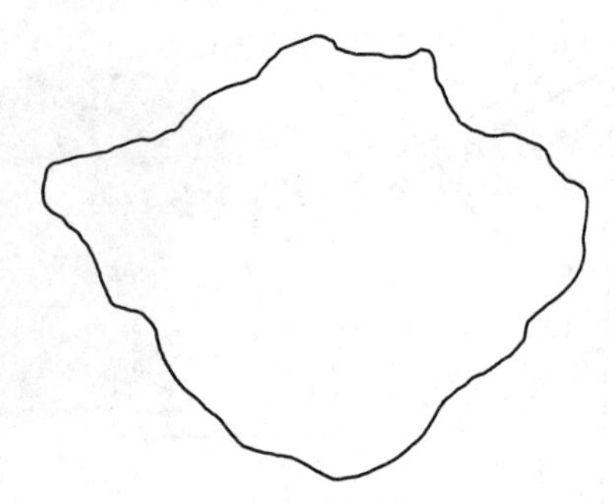

Rule a line across it to obtain one straight edge. Cut off the shaded area.

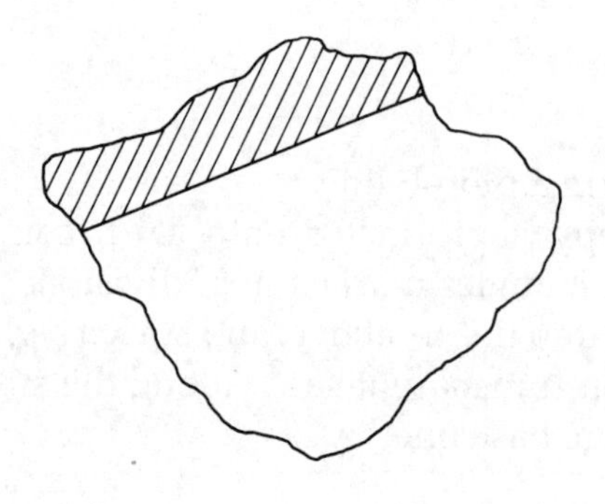

Pivot a piece of string on to the card at tube level half way along it, and put a weight, such as Blu-Tack®, on to the free end of the string. Sight the top of the tree or building and stand still.

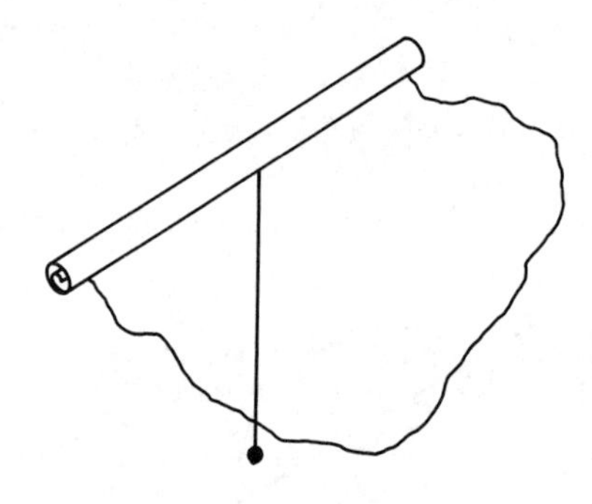

A partner tapes the end of the plumb line when it is steady at the vertical. This 'freezes', the angle. Complete other site measurements as necessary.

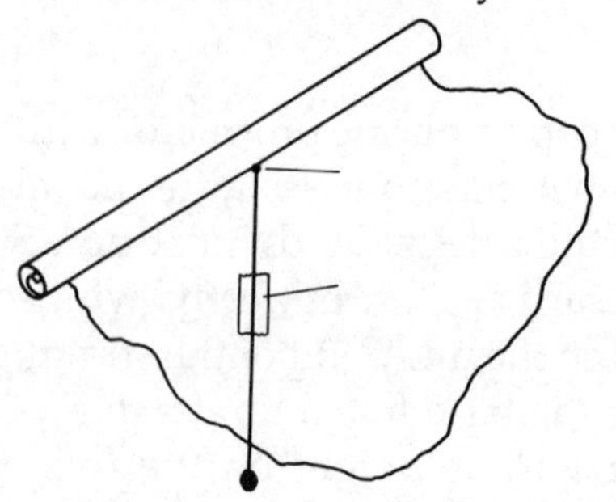

Back in the classroom, cut out the angle 'h' shape.

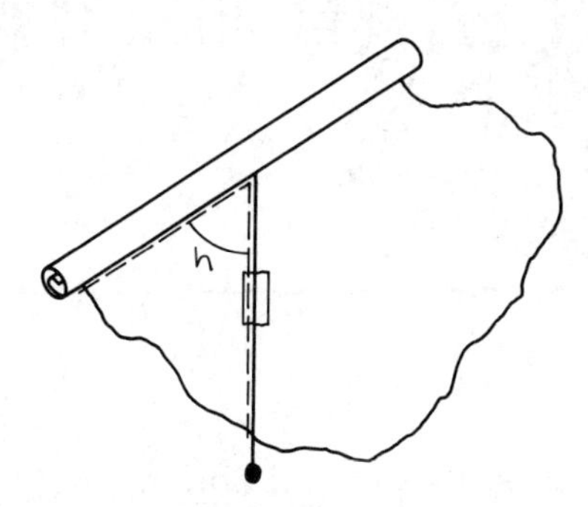

Either by folding along side *a* or by using a set square, obtain angle *e* and cut off the unwanted shaded section.

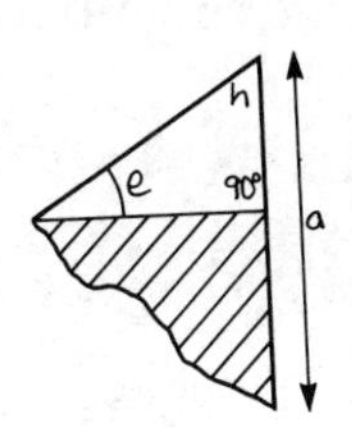

Using this angle like a template to mark the angle of elevation on squared paper. Draw on a scale such as 1 square = 1m to obtain the height of the object sighted, as explained in Figure 9.7.

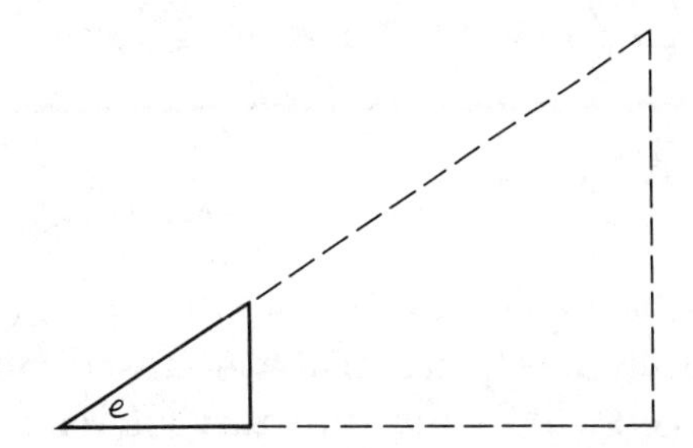

10 STUDYING PLACES

Of all the aspects of National Curriculum geography, it is 'place' that has caused most controversy. It is the difficulty of realistically assessing knowledge of places which is particularly problematic.

In Chapter 1, good practice in geography is described as much wider than the 'capes and bays approach' – the ability to name and locate places on the earth's surface. In Chapter 2, the authors noted that HMI (1989) found that primary geography demonstrated:

- Little work beyond the child's local area
- Limited work on other places in the British Isles
- Not enough use of atlases and globes.

It would seem that, although good work was being done relating to the local area, not enough was being done relating to distant places in either the narrow 'capes and bays' way or using the wider approach of good practice.

By 1995, the situation had certainly improved, and the introduction of National Curriculum Geography had certainly been a trigger for that improvement.

OFSTED was able to comment in its *Review of Inspection Findings 1993/94* that many schools provided a structured programme of local and distant visits, some residential, during which pupils frequently achieved good standards of work.

Nevertheless, it was still able to point out that the study of distant places was insufficiently thorough and did not integrate themes and skills in planning. There is still much progress to be made in achieving good practice in studying places.

Balancing local and distant places

Good practice in teaching about places should balance a scale of development from local to global. The younger the children, the more they need experience of the local and therefore visible area. This does not mean that we should deprive them of learning about distant places because they cannot experience them by direct or first-hand experience. Breadth and balance must be the key to understanding places in geography. Children need to study local and distant places in order to develop a 'sense of place'. As this grows, their local, regional and country knowledge becomes a yardstick with which to compare the rest of the world.

When drawing up the first Statutory Order, the Geography Working Group was given the thankless task of trying to produce a broad and balanced 'place' framework for children aged 5 to 16. This framework was originally very specific as to the particular countries to be studied. Great protest greeted this prescription, which has been much modified. Teachers objected to being told which places to study. We can now have the option to choose which specific places we study: it is the type of place or general location of the place which have been prescribed. You may wish to refer back to the localities, regions and countries chart in Chapter 2, Figure 2.1 at this point.

The meaning of locality

Crucial to the type of place we need to study with children is an understanding of the word 'locality.' This means a focus, like that of a camera zoom lens, on a particular place in the country. That place covers a 'small area with distinctive features'. This could mean a village, a small area of the countryside, a suburb, part of a large suburb, or a small town.

A locality does *not* mean a whole country, and it is important that we recognise the significance of this. There is a great risk that the specification in the Order to choose localities could concern us so much with the 'zoom lens' that we could forget to position these snapshots in their wider country context. Our children could end up with many detailed impressions of specific localities without ever relating them to the country or area in which they are located. It is essential that we always help children to find the 'wide-angle lens view', as well. They need to know where the locality they are studying is situated in its country or area and how it links both to other places in that country or area and to where the pupil lives. This will, of course, be common-sense good practice to many colleagues, but to others who have not dealt with geography in the context of places very much, it needs reinforcing.

The Revised Order tries to make this very clear, both in key stage 1 and 2 programmes of study.

In key stage 1, the general summary paragraph, which contains the essence of the other parts of the programme of study, states that:

> "*KS1 PoS 1 pupils should be given opportunities to: become aware that the world extends beyond their own locality, both within and outside the United Kingdom, and that the places they study exist within this broader geographical context, e.g. within a town, a region, a county.*"
>
> DFE, *Geography in the National Curriculum* (HMSO, 1995)

In key stage 2, it is emphasised even more in both the general paragraph and in the places section:

> "*KS2 PoS 1 pupils should be given opportunities to: become aware of how places fit into a wider geographical context, e.g. links within a town, a rural area, a region.*"

and

> "*KS2 PoS 5e how the localities are set within a broader geographical context, e.g. within a town, a region, a country, and are linked with other places, e.g. through the supply of goods, movement of people.*"

These points will be returned to later in the context of each locality type.

Studying the local area

Include the school

The locality of the school is a constant locality to be studied throughout key stages 1 and 2. In key stage 1 it is defined as 'the immediate vicinity: it includes the school buildings and grounds and the surrounding area within easy access' (PoS 4). In key stage 2 it should cover 'an area larger than the school's immediate vicinity. It will normally contain the homes of the majority of pupils in the school' (PoS 4).

This is consistent with good practice in that the school buildings and grounds do form part of our local area. Indeed, infant teachers always start with the geography of the classroom and school building because, until the children are familiar with those localities and can find their way about their classroom, further learning in any subject will be limited.

'Do we have to include the local area every year?' is a question teachers often ask when they plan a key stage. There cannot be a legal requirement to include the local area *every* year, but good practice indicates that teachers want to use the classroom, the school building and the area around the school as a place resource some time in each year. The amount of time for which they do so should depend on the type of geography work in the topics being undertaken, which programme of study aspects are being developed and on knowledge of what children have done before in these areas, assuming good record-keeping has been passed on. Progression in the activities is crucial in both fieldwork and map skills as explained in Chapters 8 and 9. Progression in the depth of physical and human geography environment study in these areas is also crucial. For vertically grouped schools it is likely that classroom, school building, school grounds, and local area will all have to be utilised in every class to secure adequate differentiation. For horizontally grouped classes it may only be possible to go beyond the school gates once in the infants, once in the lower juniors and once in the upper juniors. The authors would, however, like to stress that children should preferably be actively involved in investigating their local environment every year.

Using the classroom

The comparatively safe and limited spatial locality of the classroom has most potential for infants and for reinforcement activities for older, low ability learners.

Concepts relating to human geography and mapwork skills are the most useful areas of learning to be developed here.

The following suggestions for using the classroom locality may be useful.

- Children can locate 'services' in the classroom – Where are the scissors, paper, maths equipment and water supply kept? Children can describe the locations with geographical language or map them on a plan drawn by the teacher.
- Recognising the teacher's signs and symbols used to label these services is part of developing representational skills.
- Children can play 'signpost' mapping games (see Chapter 8).
- Children can plot their route around the classroom on their own maps or on the teacher's base map.
- Children can talk about the size and shape of the classroom compared to other classrooms or their rooms at home.
- Children can begin distance work with a scale drawing of the classroom (as in Chapter 8). Carpet or floor tiles can be useful here for matching relative position and scale. If the classroom size is '10 rows' by '15 columns' of carpet tiles, then teacher or children can mark this off on squared paper or an overhead projector transparency grid and draw or position cut-out furniture, and so on, accordingly.
- Younger children can make sketch maps of their classroom, trying to get relative distance and positioning correct.

Using the school buildings

Again, this has greatest potential for key

stage 1 pupils, although those children transferring to a separate junior school or to a different building on the same site will benefit from a quick revision and extension of activities already done at infant stage. They will need quickly to establish their new sense of place and space in the alternative environment.

The following suggestions for using the school buildings may be useful.

- Children can make up a trail inside the school building. They can then follow it themselves to check it, and ask another pair of children to follow it, too. Does the trail work or does it need improving? It is a word trail, a photo trail, a map trail, or a puzzle trail? Can children escort visitors or other children to locations in the school? These obvious and important confidence-building activities are very geographical! Trails are referred to in more detail on page 128.
- Lay a series of signs around the school to develop a trail following and understanding signs and symbols: cut-out footsteps, arrows, etc. which will stick to the floor without damage, pictures on a theme placed along the walls.
- Can the children do a routeway survey in the school? Can they map their results? Can they make recommendations in concluding, for example, which are the most used corridors?
- Explore the school as a workplace. Who works here? Where do they work? Map the results to record them on a plan of the school with colour codes. How many types of job are there? For example, caretaking; cleaning; teaching; classroom assistant; food preparation work; administrative work. Are they service jobs – do they help other people? Are the industrial jobs producing goods?
- How is the building itself used? Can children map it according to use or function? Use a colour key or picture symbols to show, for example, the kitchen areas, caretaker's room, administrative areas, teaching/learning areas. This is elementary work on land use.
- With older juniors, address microclimate within the building. Is one part of the building colder and damper than other parts? Carry out an enquiry to find out by measuring temperature, aspect (which direction the rooms face), trying to work out whether the walls are cavity walls or not. Is a room colder because of its location: because it is situated furthest away from the boiler; because of aspect, window size, poor insulation, or what?
- With children in school buildings of manageable size and shape, one-storey and flat roofs, can a scale model of the buildings be made to be fitted on a scale plan base of the grounds and building? Older children can make such a model to give to the infant department to develop plan view work.

Using the school grounds

The school grounds are a superb resource for geographical work whose potential is still not fully explored in spite of the arrival of National Curriculum geography. School grounds are often extensively used for environmental science (Sc AT2), but less for geography. Schools with fields and trees in their grounds will clearly be able to make wider use of their grounds than those schools with playground areas of concrete or tarmac only. Nevertheless, even the school which feels it has the most unstimulating grounds can deal with all aspects of school grounds geography – physical, human and environmental, except for soil and natural vegetation analysis and mapping.

The following suggestions for using the school grounds may be useful. First, let us look at the activities that can be done in the school grounds.

- Figure 10.1 indicates the wide range of activities possible to develop geography in the physical and human environments.
- All the mapping and fieldwork skills and techniques included in Chapters 8 and 9 can be used in the context of the school grounds to develop physical, human or environmental geography themes, as well.
- Many enquiry process activities, including some of those listed in Chapter 1, can be undertaken in the school grounds, for example:
  - Where is the best place to site the new flower bed/seats/trees/compass rose drawn on the playground?
  - Are the school grounds and buildings a secure place?
  - How can we improve our grounds?
  - Can we make a trail for younger children to follow?
  - Where were those photographs taken from?

Second, let us look at the advantages of activities in the school grounds.

- Problems of supervision are alleviated.
- Often it is possible to set the children to work in groups within view of the teacher who is still able to work inside with the rest of the class.
- Flexibility in timing the work is possible.
- The environment is a more limited, safe and controllable one.
- Skills can be safely and simply practised in a secure geographical context *before* being applied in a new and distant environment.

To ensure no overlap or unnecessary repetition, continuity and progression in the use of the school grounds as well as the classroom and buildings needs to be established by the geography coordinator and the staff by planning opportunities from the PoS for key stages 1 and 2, as suggested in Chapter 4.

Don't forget that many school grounds offer the opportunity to begin work on the vicinity of the school. In schools situated in a valley, whether urban or rural, or on a hilltop or slope, views out beyond the school grounds may offer the opportunity to do safe and easy work on the local area. In schools not hemmed in by high walls or other barriers over which children cannot see, just a walk around the perimeter of the grounds may enable land use and building form and function to be observed. Changes can also be observed through the fence (see Figure 10.2).

The local area: beyond the school

As previously mentioned (page 119), the Revised Order defines the locality in a slightly broader way for key stage 2 than for key stage 1. Nevertheless, it should still be a convenient area in which you can organise fieldwork, as suggested in Chapter 9, Figure 9.1, at least once in every local area unit of work. It is assumed that children's fieldwork and background knowledge will enable them to investigate this locality. We need to remember that fieldwork can be crucial for many children, as mobility is such that they may travel daily many miles by car to their school or be boarders and so have no 'sense of place' about the school locality.

Teachers, too, may commute many miles to work in a school about whose local area they have little knowledge, so it is important that the geography coordinator encour-

Figure 10.1

<table>
<tr><th colspan="3">Use of the school grounds from a geographical perspective</th></tr>
<tr><td colspan="3">Best use is made of any environment if the context is motivating, e.g. enquiry or issue-based.
Some possible contexts: children need to know the best place to site new trees, tubs of flowers, compass rose, direction sign to office, KS2 children planning school grounds activities for KS1 children to use, etc.
Map-making and using skills can be used throughout this work in context – see Figure 8.1</td></tr>
<tr><td>What natural environment work can we address?</td><td>What human environment work can we address?</td><td>What specifically environmental issues can we address?</td></tr>
<tr><td>Physical geography

1* Soils and rocks
Take soil samples and analyse. Compare to children's garden samples, samples taken from locality near school, etc.
2 Weather and climate
Basic weather and seasonal change observation – including cloud cover and type.
Micro climate studies – sun and shade site study – contrast in temperature, open aspect to wind, shelter from rain, etc.
3* Slope and run off (water) i.e. drainage
Does the water from rain gather in certain places? Does it take longer to evaporate in some sites than others?
4 Vegetation
Land use mapping.
Transects, flower bed mapping.
Quadrats, vegetation mapping.

* There are overlaps with National Curriculum Science here.</td><td>Human geography

1 Built environment
School building, walls, playground: Natural or man made materials? Identification/classification.
2 Human traffic
Examine traffic around the school site. Conduct surveys. Map pressure points.
Are there zones of greater/lesser use? Can they be mapped?
Concept of shortest route/best route. Are 'official' routes the most direct or are the 'unofficial' ones shortest?
Concept of barriers – detour linked with best route. Where are the barriers – steps, fences, car park areas, forbidden areas?
3 Function
Map the function of different parts of the school, e.g. car park, office, etc.
4 Services
Identify and trace service points around the school – telephone cables, drains, gas, electricity entry points. Safety issues: look, don't touch.</td><td>1 Air pollution
Place filter paper in funnel of bottle – leave for a certain time. Test rain water and tree bark for acidity. Test for lichen and algae growth.
2 Noise pollution
Develop a scale for testing noise pollution – lorries passing by, voices, aeroplanes, etc.
3 Visual pollution
Mapping of litter or vandalism pollution, etc.
4 Environmental improvement
Is there some project we can undertake to improve the grounds?</td></tr>
</table>

ages staff to share knowledge about the local area and to come to a consensus about the suitability of various sites in it. This could be done in a staff meeting or as an INSET activity. Figure 10.3 provides a list of the kind of sites and features you will need to think about and plan to use in your local area for children at some stage between ages 5 and 11.

Common sense and the 1995 Revised Order tell us that sometimes we may need to extend the definition of the local area to something a little larger than the immediate vicinity of the school. Here are some examples of when this may be sensible:

- Is your locality a very small village with limited features? Maybe you need to embrace a larger village or a nearby small town with which your village has strong links
- Is your locality very flat and featureless, or uniform, for example a very large, level housing estate? Maybe you could enlarge it to embrace the whole town area, including the town centre. In order to understand their locality the children need to see it in the context of the town of which it is a part.

Having established the potentially flexible extent of our local area, we can then plan in

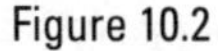

Figure 10.2

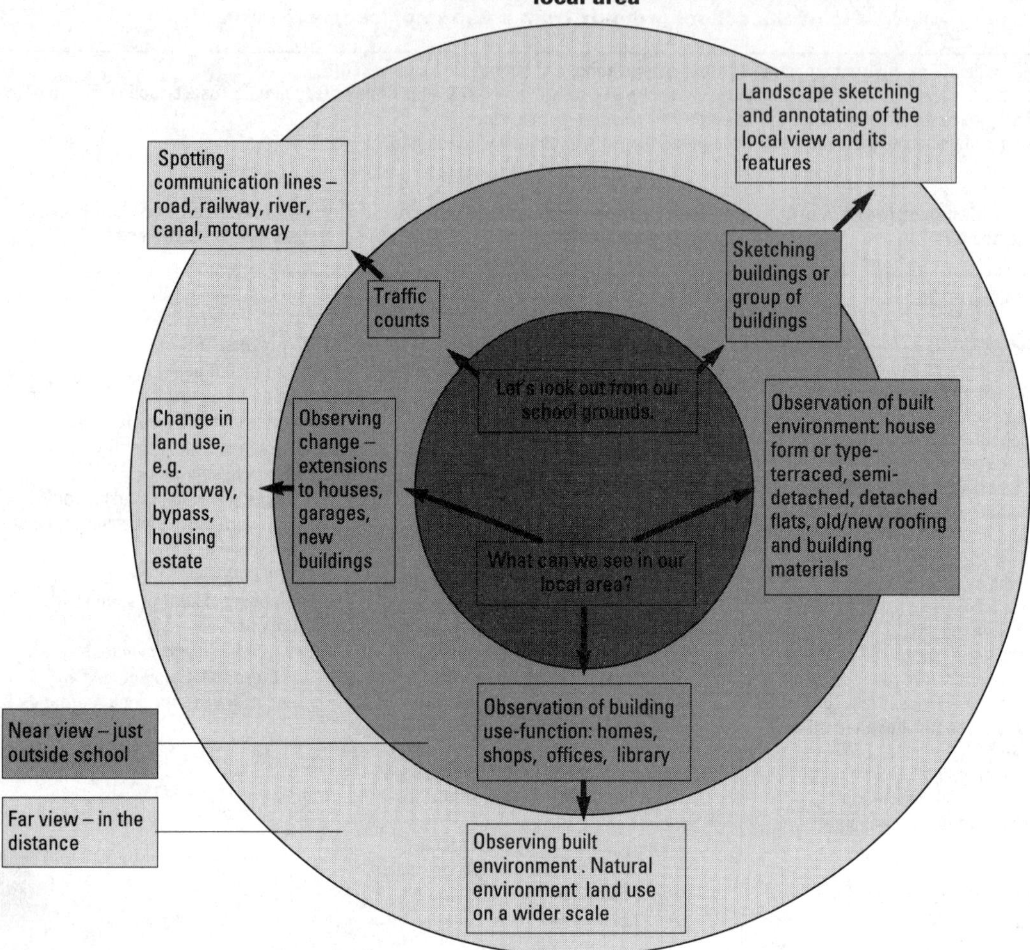

more detail which physical and human geography we will study with our children across the two key stages. Environmental geography should arise through the human and physical themes and local issues. Planning units of work on the local area should revolve around the seven key questions shown in Figure 10.4, whether the units are purely geographical or form a smaller or minor part of a larger multi-subject-focused topic.

Asking these questions help focus on the subject matter of geography. You will want to focus on the concepts behind settlement or the local landscape, for example, depending on which part of the programme of study you are developing in your units of work, and how they fit with other subjects.

Progression in the local area

Teachers often ask for a model of study in the local area to show progression. As each locality is different and will have different resources, it is, unfortunately, not possible to provide such a model, but some guidelines may be useful.

Figure 10.3

Checking up on your school's locality

Which of these features do you have?
Ideally you need at least one of each type for local area work for pupils ages 5–11

Water features
Stream, pond or lake
River
Estuary
Coastal area

Landscape features
Hills, valleys, cliffs, mountains, which show evidence of erosion or deposition by water, wind or ice; vegetation

Physical features
for slopes, soil, rocks

Climate work sites
for weather surveys, micro-climate work (usually school grounds)

Local issues
By pass?
Road widening scheme?
Out-of-town shopping development?
New housing estate?
New reservoir?
Rubbish tip site?
Local improvement scheme?

Sites showing the origins of settlement
Crossing point of a river
A route centre
A defensive site
Site where water became available
Old core of modern settlement
Beginnings – growth/development – decline evidence

Buildings
House
Row of houses
Housing estate
Groups, rows of buildings to examine function

Transport
Safe place for traffic survey
Bus station
By-pass
Airport
Railway station

Industry
Farm
Business
Small manufacturing unit
Warehouse
Factory

Shops
Single shop
Parade of shops
Supermarket
Hypermarket
Shopping mall

Leisure facilities
Library
Park
Swimming pool
Leisure centre
Golf course

Settlements
House
Village
Town
Suburb
City – suburb

Services
Fire
Police
Ambulance
Hospital
Doctors
Refuse collection
Recycling plant

- Beware of the large and complex environment. The larger the settlement, or area of a settlement, the richer yet more confusing it will be for the children. Your local row of shops may be less exciting than studying the Shambles in York or a hypermarket, but it is much more manageable for you and the pupils.
- The smaller the settlement, the easier it is for the children to get an overview of what goes on there.
- The younger the children, the fewer the features it is advisable to investigate at any one time. Form and function – what type, size and age the buildings are, and what they are used for – is a good illustration of this:
  - Top infants investigating their local homes could focus only on about three houses in a row. To begin with, it could be enough for them to identify what the houses seem to be built of, whether they are lived in or not, and the different colours of the houses' front doors. As both you and your children become more experienced, the enquiry can broaden
  - Lower juniors may consider a limited row of local shops – say six – and be asked to identify what kind of shops they are and note this on a large teacher-prepared map of an enlarged section of an OS map
  - Upper juniors may be divided into small groups, with each group being required to assess a defined section of a street according to ground-floor, first-floor and any further floor's use. Their exercise may be in the context of an enquiry such as 'How has our High

Figure 10.4

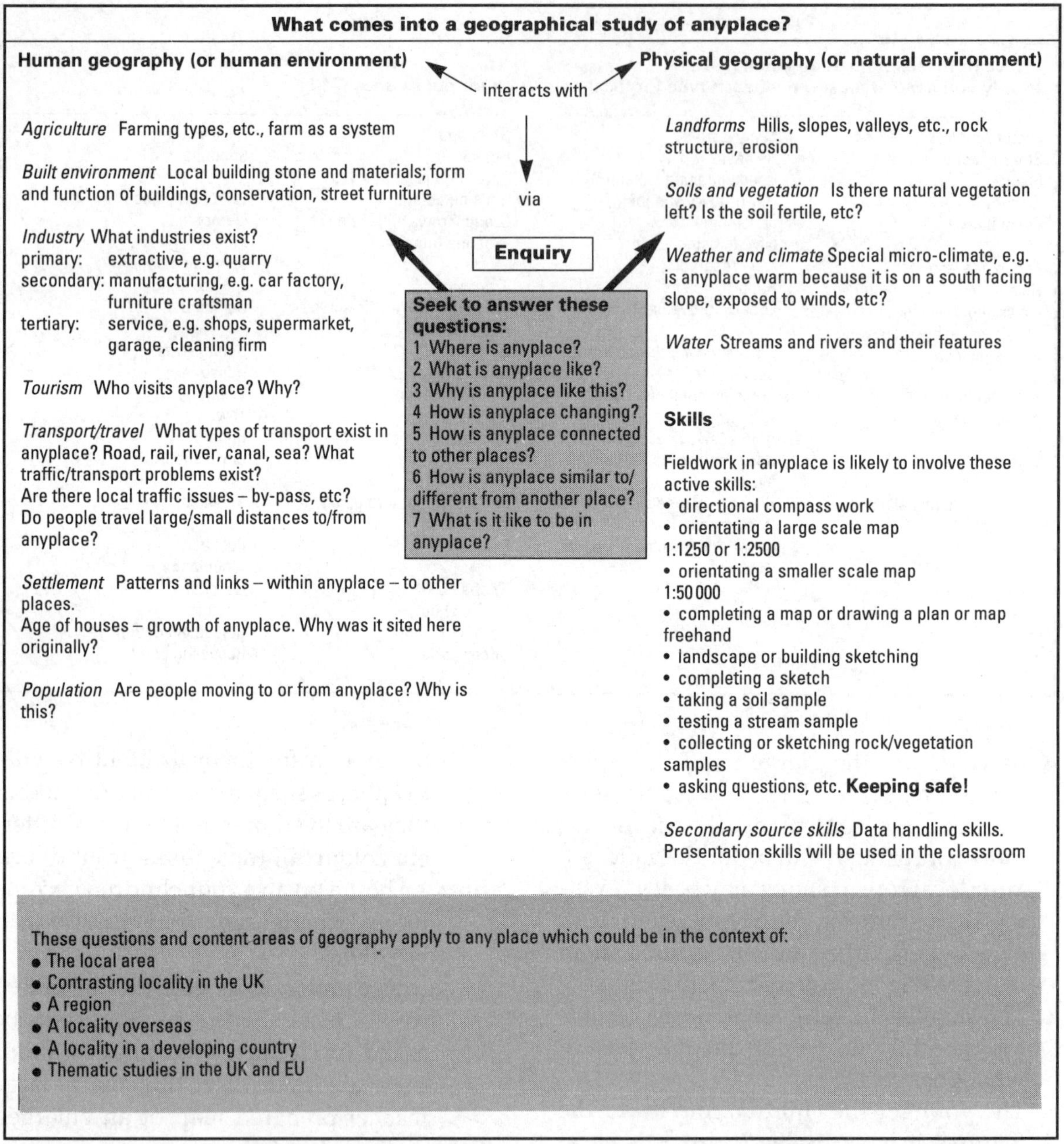

What comes into a geographical study of anyplace?

Human geography (or human environment) ← interacts with → **Physical geography (or natural environment)**

↓ via

Enquiry

Seek to answer these questions:
1 Where is anyplace?
2 What is anyplace like?
3 Why is anyplace like this?
4 How is anyplace changing?
5 How is anyplace connected to other places?
6 How is anyplace similar to/different from another place?
7 What is it like to be in anyplace?

Agriculture Farming types, etc., farm as a system

Built environment Local building stone and materials; form and function of buildings, conservation, street furniture

Industry What industries exist?
primary: extractive, e.g. quarry
secondary: manufacturing, e.g. car factory, furniture craftsman
tertiary: service, e.g. shops, supermarket, garage, cleaning firm

Tourism Who visits anyplace? Why?

Transport/travel What types of transport exist in anyplace? Road, rail, river, canal, sea? What traffic/transport problems exist?
Are there local traffic issues – by-pass, etc?
Do people travel large/small distances to/from anyplace?

Settlement Patterns and links – within anyplace – to other places.
Age of houses – growth of anyplace. Why was it sited here originally?

Population Are people moving to or from anyplace? Why is this?

Landforms Hills, slopes, valleys, etc., rock structure, erosion

Soils and vegetation Is there natural vegetation left? Is the soil fertile, etc?

Weather and climate Special micro-climate, e.g. is anyplace warm because it is on a south facing slope, exposed to winds, etc?

Water Streams and rivers and their features

Skills

Fieldwork in anyplace is likely to involve these active skills:
- directional compass work
- orientating a large scale map 1:1250 or 1:2500
- orientating a smaller scale map 1:50 000
- completing a map or drawing a plan or map freehand
- landscape or building sketching
- completing a sketch
- taking a soil sample
- testing a stream sample
- collecting or sketching rock/vegetation samples
- asking questions, etc. **Keeping safe!**

Secondary source skills Data handling skills. Presentation skills will be used in the classroom

These questions and content areas of geography apply to any place which could be in the context of:
- The local area
- Contrasting locality in the UK
- A region
- A locality overseas
- A locality in a developing country
- Thematic studies in the UK and EU

Street changed since 1840 in Victorian times?' whereas the infants may be answering the more limited key question 'What are the houses like around our school?'

– The younger the children, the more their geographical activities should be concerned with basic *observation* and *recognition* of features, with recording being secondary, although some data-recording can be done by the more able, or on a rotating basis.

- The older the child, the more observation

and recognition should be speedy and used as a means of noticing changes, similarities and differences, and recording data in map or note form, possibly using IT.

A suggestion for starting local area work

Most teachers and parents want children to know where they live in a practical way – by being able to recite their address and by having a concept of how to find their own home in the real environment.

The following series of activities could be used to start off key stage 1 work in the local area in the context of a geographical or cross-curricular topic on 'My House', 'Homes' or 'My Family.'

These activities, done by Kent teacher Sue Thomas with her class, turned 'Know your address', which could so easily be a boring, rote-learning exercise, into a constructive and active learning experience which children loved.

1 Children are asked to observe and sketch their front door at home. A discussion on 'My front door' then leads the children to talk about the materials, colours and fittings. As well as the design features they notice, some children will volunteer that their door has a name or a number.

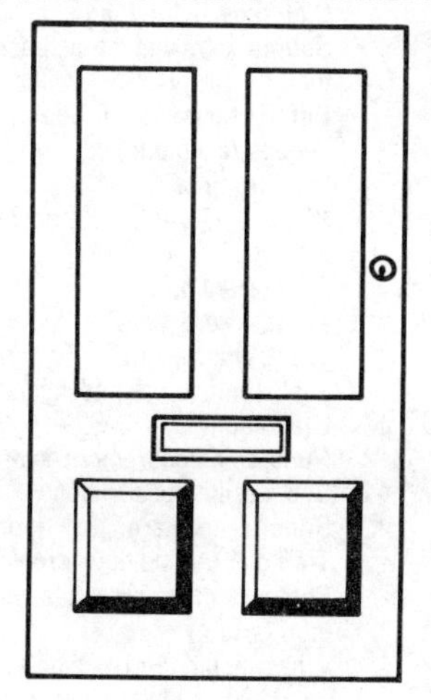

A further discussion on the various names and numbers of their houses lead to the reasons for identifying houses in this way. Each child can be given a numbered card to hold, and the class can arrange itself into a road, with odd numbers on one side and evens on the other. Other combinations can be tried out to simulate cul-de-sacs and 'walks.' Some children can act as post and milk delivery people, and show their delivery routes. Follow-up work could include drawing their door with a name or number and drawing a street with the post person's route marked.

2 The second session can be used to discuss all the different words used to describe roads – What is a lane, a close or a crescent? Photographs of different types of streets are a useful resource to show children that different places have different patterns of houses. A classroom display of the street names where each child lives is a useful follow-up activity.

Eldorado Road

3 From a starting point of whether we live in a village, town or city, a discussion on the characteristics of these three types of place will lead the children to talk about the names of their village, town or city and of nearby villages, towns or cities. They are able to talk about the differences in size of local places and also which way you have to go to reach them.

4 Making a signpost to local places and standing it up in the classroom or playground will help to strengthen this concept, and is a good introduction to work on direction.

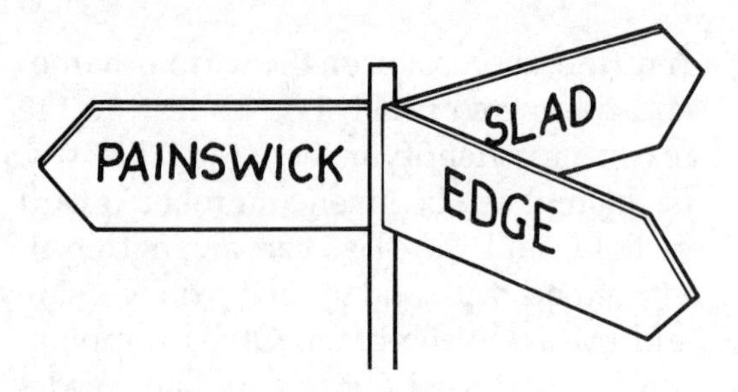

After these sessions, the children are ready to draw their house and to write or copy their address beside it, knowing the components and what each one means, and also understanding the usefulness of knowing people's addresses.

Further work can include making a poster for a lost dog which includes the owner's name and address, and addressing envelopes for special cards taken home after an art session, for example Mother's Day, Easter and Christmas.

Local trails

Developing local trails is a tried and tested way of encouraging geographical, environmental, and cross-curricular work. Teachers can develop them, but a good enquiry exercise is to get a class to develop one for other classes to use and for your own next year's class to refine. The aims and objectives of local trail work should be as shown overleaf.

1. To extend the children's knowledge about their own locality
2. To enable the children to develop the ideas of:
   - Location
   - Distribution
   - Networks
   - Distance
   - Scale
   - Similarity and difference
   - Comparison
   - Continuity and change
   - Cause and effect
   - Time.
3. To develop skills: written, spoken, numerical and graphical
4. To increase children's motivation and involvement through the use of first-hand experience
5. To encourage children to observe and be investigative
6. To increase environmental awareness by fostering an interest and concern for the quality of their local area.

Figure 10.5 suggests some ideas for use on local trails. Can you focus on these ideas and change them, or ask children to change them, into an enquiry to build up a trail?

Remember the following:

- Is the route safe?
- Variety in recording methods
- Green Cross Code
- Different eye levels. Look up, through, under or over

Figure 10.5

Local trails	
Houses	Materials, doors, windows, types
Boundaries/ barriers	Fences, hedges, walls, council, safety
Underground	Gas, water, telephone, sewers
Patterns	Around, on different scales
Change	Over time, on-going
Development of the area	Over time, in the future
Pollution/litter	Source, cars, industrial, possible solutions
Trees	Different times of the year
Gardens	Type, size, colour
Habitats	Insects, birds
Shops	Hierarchy: Who uses them? What do they sell?
Feeling materials	At different places
Street furniture	What, where, why?
Words	What, where, why?
Transport	Evaluation, future possibilities
Good, bad and ugly	Evaluation
Maths trails	Number, measurement, symmetry
Park, wasteland	Use, design possibilities
Design	General small area, e.g. doors
Routes	Traffic or people flow, crossing points
Spotting trails	Photographs or sketches, useful for infants
Tourism	What should tourists stop at/visit? What would you want to show visiting friends?

- Length of walk
- Arrange extra adults.

Teachers or children making up local trails may find the following tips useful.

1. See that the children are well briefed before they set out, have had a chance to read through the trail, and know what road behaviour is expected of them.
2. Arrange for parents or other adults to help supervise the class during the outing. Inform your head teacher and complete any necessary paperwork, such as consent forms. Remember the safety guidelines provided in Chapter 9.
3. Decide exactly where you are going to cross roads – even busy streets are quite safe as long as the children are given clear instructions.
4. Don't always select the obvious route. If the trail leads down paths and alleyways, it is more likely to contain surprises and contrasts.
5. Choose a circular route, if possible, as this is most convenient and avoids a lengthy walk back to school.
6. Don't make the trail too long – eight stops are ideal – and always mark them clearly on the trail map.
7. Select the stops carefully, so that each one illustrates a single, definite idea.
8. Keep any text down to a minimum, so that the children do not spend too much time reading rather than working.
9. Include a range of recording techniques, such as surveys, diagrams, annotated sketches and mapwork exercises. Beware of questions which simply test the child's knowledge.
10. If possible, use illustrations to provide extra information, for example about the changes that have happened to a building or street.

Always bear in mind the following:

1. The urban environment is always changing. Be an opportunist and take advantage of chance events if they occur!
2. In the local area, whatever the age of the child, they will need to focus on these questions to do with description and analysis:
   - Where is the place?
   - Why is it like this here?
   - How is this place changing?
   - How is this place similar to or different from other places we know?

Studying distant places

Distant places for primary school children are generally acknowledged by their teachers to be anywhere that is beyond the local area – indeed, anywhere beyond their personal experience at that moment. That is why these places can be:

- A contrasting UK locality
- A contrasting locality overseas
- Part of the broader or wider geographical context, for example within a town, a region, a county, continent or global situation.

The distant localities are included in this section of the chapter. The same principles and strategies apply to teaching about all those localities of which the children have no, or very limited, first-hand experience.

A contrasting UK locality

Studying a contrasting UK locality is an alternative for pupils in key stage 1, if an overseas locality option is not chosen. However, a UK contrasting locality is compulsory for pupils in key stage 2.

How do we define a contrasting locality in size and type?

Size The Statutory Order (1995) defines it for us in KS1 PoS 4 – Places.

In key stage 1, it should be an area of similar size to the vicinity of the school – an area in which children can make almost one to one comparisons with their own area. Notwithstanding this definition, the Order reminds us in KS1 PoS 1c that the locality must be placed within its broader geographical context, i.e. its relative position within a town, rural area, region or within the UK.

In key stage 2, the contrasting locality will be bigger than in key stage 1, covering an area larger than the immediate vicinity. For example, it will normally equate with an area the catchment size of the school.

Similarly, this locality should not just be a total 'zoom lens' approach – it will need to present the wider angle, situating the locality within its wider geographical context (KS2 PoS 1d) and showing links with other parts of the town, region or UK (KS2 PoS 5e).

Type of locality Exactly what does *contrasting* mean?

Common sense needs to rule here. Teachers are very concerned to know that their contrasting locality is sufficiently contrasting. Given that life styles within the UK demonstrate more similarity than difference, it is sometimes difficult to find resources for an appropriate locality.

What is needed is a reasonable number of geographically contrasting features in your UK locality. If you can apply some of these criteria to your contrasting locality, then it is likely to be appropriate. Here are a few examples:

- different physical features to your school's locality
  - hilly as opposed to flat
  - high and exposed in winter as opposed to low and kept mild by maritime influences
  - coastal as opposed to inland
  - very different rock and soil characteristics which develop very different landscape features which reflect in the human geography
- different human features or activities
  - quarrying as opposed to farming
  - a different kind of farming
  - a different urban environment with high-rise blocks and large shopping areas as opposed to housing estates and neighbourhood shops.

The greater the number of contrasts, the more different and therefore the more stimulating the contrasting locality is likely to be.

Schools in rural areas feel that they must choose urban localities. This is certainly the most obvious human geography contrast. However, if you have better quality resources for secondhand learning on a *contrasting* rural locality which you feel will promote better learning, then that is a justifiable option.

A village school near the sea in Kent where there is arable farming and no tourism may usefully study a contrasting tourist village in the high area of North Wales where the climate is wetter and cooler, where slate is the traditional roofing material and a former source of employment in the area, and where hill farming is different from that of the arable downland.

The same may be the case for urban localities. A school in suburban Plymouth in the mild, coastal south-west of the UK may find sufficient contrasts in an inner-city area of Sheffield with its cultural contrasts, more severe winters, different industries and inland situation which makes it an important route centre in its area.

The Revised Order (1995) is prescriptive in that the locality in the UK must be 'contrasting'. It allows for considerable freedom in common sense geographical interpretation by not defining just which criteria or how many must be used in deciding on a locality. This is part of the freedom of the Revised Order, highlighted at the end of Chapter 2.

In considering breadth and balance in places over key stages 1 and 2, if the contrasting UK locality option is taken up in key stage 1 as opposed to any overseas locality, then it makes sense to choose a different kind of environment in key stage 2.

The world *is* a big place and the more experience children can gain of it through first-hand or vicarious experience the better.

The various places to be learned about nest within each other as shown in Figure 10.6. The local area work that has taken place in infant and lower junior levels is still ongoing. Children need to locate their local area on maps in relation to their surrounding region – wall maps, UK atlas maps, regional maps – to help reinforce the notion that where they are is part of a wider area. This reinforces the 'zoom lens' altering to 'wide-angle lens' focus which is so important.

Every time children leave their local area for some fieldwork, be it a visit related to science, history, geography, or an arts visit to the theatre, they should locate where they are going on a map.

Teachers should draw on those visits made in the surrounding region from infant age onwards to remind children that they have many 'snapshots' of their home region which can help piece together the different features and activities which go on there. An idea would be for the school to have a permanent local region map in a central place where label flags are planted for every visit which occurs.

Figure 10.6

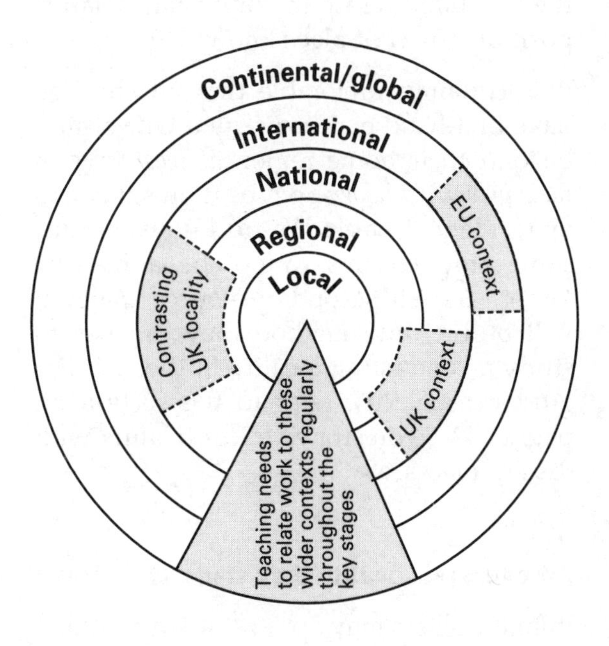

Teachers will also have to rely on the children's knowledge to build up wider context work related to where they live – an acknowledged problem where mobility and family outings are limited. Involve the class in questionnaires and map the results to get a picture of their knowledge and experience: Where do you go to the seaside for the day? Which is the biggest shopping centre you go to?

In key stage 2, the key question and enquiry process approach is the most successful way to motivate work on the wider context of the school locality, along with investigating issues: What will the effects of building the motorway extension be? Where will it go to? Why is the superstore being built and what effects will it have on people living here?

Don't forget to relate the position of the home region to its position within the rest of the UK, building up the important relative position aspect of place knowledge.

It is certainly preferable that all children have first-hand experience in a UK locality beyond their home region *at least once in their primary school career*, be it on a day trip or residential basis. We do know that for some children this cannot be a realistic target, and so secondary source material will be the only method they can use to study a contrasting locality in the UK. For guidance on this, refer to the section on pages 142–50 on strategies for dealing with distant places.

An overseas locality (key stage 1)

Infant teachers may opt to study a contrasting locality overseas, instead of a contrasting UK one. Again such a locality should be about the size of the school locality and have physical and/or human features which contrast with those in the school's locality.

Infant staff have the advantage of being able to choose any locality in the world to study a camera-zoom lens picture. They can, therefore, if they wish, draw upon their own or their pupils' experiences and resources, if they choose a locality beyond the UK, as well as published resources. The essential factor is that sufficient resources should exist and be available to the children to bring the plan to life for them. For guidance on this, refer to page 181 in Chapter 13 (Resources).

A locality in an economically developing country (key stage 2)

'How do we define an economically developing country?' most teachers ask. By common sense, is the pragmatic answer.

Technically, an 'economically developing country' is one whose gross national product falls below a certain figure in a table of statistics. The UK is considered to be an economically developed country, so it makes sense for children to study places with a contrasting lifestyle, which are not as developed. Values and attitudes are bound to surface in any such study and an open-minded approach is essential when comparing and contrasting 'developed' and 'developing' localities and countries.

The first Statutory Order did not define the term 'economically developing' and, therefore, left this key stage 2 locality open to interpretation. However, the Revised Order is precise in locating this contrasting locality for key stage 2. It must be in a country in Africa, Asia (excluding Japan), South America or Central America, including the Caribbean (KS2 PoS 4). Plenty of scope for choice! Once again the key to developing appropriate work for juniors is sufficient appropriate resources and knowledge (see pages 181–6, Chapter 13). Teachers' personal preferences or the linking with the Aztecs in Mexico as the modern contrast to the history study unit, where few effective resources exist at locality level, are not adequate reasons for selecting a locality in Mexico, for example.

A school may decide that its economically developing locality will be in Tunisia or Egypt because it has resources on such localities – this is then a valid criterion for choice.

The advice is then, to choose your own economically developing locality, according to resources and integration with other subjects, for example a history unit on Egypt where sufficient locality resources on modern Cairo do now exist (see Chapter 13).

The European Union

> "*KS2 PoS 6 ... contexts (relating to themes and places) should include the European Union.*"
>
> DFE, *Geography in the Nation Curriculum* (HMSO, 1995)

If the European Union is a context for places and themes, then the Revised Order gives teachers immense freedom as to what they do about the European Union, and when. However, it must be dealt with, and it often seems to be left out in geography curriculum planning, because it does not appear as a specific place in PoS 4, but hides away at the end of KS2 PoS paragraph 6.

Possible EU scenarios:

- A locality study in an EU country in Year 5 looking at themes there, in a blocked unit
- A study of an EU country in Year 6 through some of the themes as appropriate, in a blocked unit
- A policy to study the EU through topical examples in every year, as a continuous unit, as geographical topics are highlighted on the news, for example weather, environmental issues
- From the UK to France and other EU destinations by Eurostar and TGV. Year 5, as a linked unit with maths investigations on distance, time and costing
- Some important rivers in EU countries in connection with PoS 7 (rivers), PoS 8b and c (flooding) and PoS 9 aspects of a, b settlement along those rivers – a blocked unit
- A study of several, or a range of, localities and their contexts in EU countries, investigating settlement, weather and environmental themes, as the 1995/6 Landmarks unit does.

Some aspects of the EU may be done in every year, or it could be confined in depth to one year. However, for the purposes of continuity and progression, references to the EU countries should be part of cross-curricular work in an on-going way, to assist pupils build up a spatial mental image of the map of Europe with its points of reference by the end of key stage 2.

Whichever way an EU place is incorporated into a scheme of work, that place should always be situated within *its* wider locality and its links with the UK and the pupils' UK area shown via the use of atlases, real maps of the foreign country and, if possible, pupils' own experiences, as many have travelled to other parts of Europe.

The United Kingdom context

This appears clearly as a place context for key stage 2, like the European Union. It is likely that by the beginning of key stage 2, pupils, because of directions in the key stage 1 programme of study, will have been made aware of the countries of the UK and have had some experience of the wider UK context, through the need to develop locational knowledge about it.

In key stage 2, teachers may plan to deal with the UK context as they wish – through themes, places, the wider context for contrasting UK localities and the local area and/or a combination of several of these.

Be wary of teaching about the UK as an entity, or as a distinct unit of work. Unless you have a range of very stimulating resources, such learning is likely to be too abstract for pupils at this age.

It is better to use a national scale example or an example somewhere else in the UK to illustrate theme work. For example, look at a nationally important river after doing practical work on a local stream or river earlier in a term or earlier in the key stage.

The UK needs to be almost a continuing unit of work throughout key stage 2 – opportunities should be taken to increase locational knowledge and understanding about its physical and human geography patterns and processes for some time, albeit often briefly in every year. Topical events should also be used as a context to do this, but they need recording at the end of the medium-term plan and passing on to the next year so that there is awareness of coverage and progression can be continued.

The broad geographical context

As already mentioned, the Revised Order is very concerned that children as young learners should be able to situate these various contrasting localities within their wider area and relate them to other parts of their region, country or continent, as appropriate. The references in the programmes of study for both key stages – in the summary part of PoS (1), the place part (5), and the skills part (3), where using globes, maps and plans at a variety of scales can also be part of this 'wider context' work – make this very clear.

With local work, it is always a good idea to plan in opportunities to develop the children's concepts of the wider world within which their own locality exists, starting from their own experience. The following work done with Year 2 classes at the New Ash Green Infants' School shows how well this can be done. Pupils follow a scheme of work which identifies blocked and linked units of work for geography and other foundation subjects under topic headings. There is an in-built progression of local area work from Reception onwards, which focuses on the school, its buildings and the local large and partly purpose-built village. Pupils have already focused on oblique aerial views of the school, identified their homes and services in the village, and mapped routes around the village, using and making maps. Teachers at an appropriate point, as part of a reinforcing of children's addresses, knowledge of the UK and the position of New Ash Green within it, then stimulate the children through discussion to think about how distant places relate to where they live on the following continuum:

my home → places very near my home → places near my home → places a long way from my home.

Definitions of these items are discussed, for they will, of course, mean different things to different Year 2 pupils, depending on their spatial ability, maturity and personal experience. Different examples which could fall into the varying categories, now defined diagrammatically in concentric zones for the children (see Figure 10.7), are discussed with children keeping their own personal examples in the chosen zones as long as they can justify their reasons. The special needs group, working with a special needs support teacher, work as a group and contribute to a large wall display concentric circle diagram, dividing the work between them and contributing drawings of places in the different zones. Other pupils fill in the concentric rings with their pictorial and/or part plan view drawings and labels, working individually. The range of different outcomes are appropriate and informative. A child who lives on an isolated farm and can only travel by car rightly perceives the infant school as located in a more distant zone than the nearer junior school, passed on the way (Figure 10.7). His own experience leads him to see Alton Towers as very distant, whereas another better travelled child who has been to Ireland sees this country and Bosnia, heard about through the media, as the limit of her world (Figure 10.8)

Figure 10.7

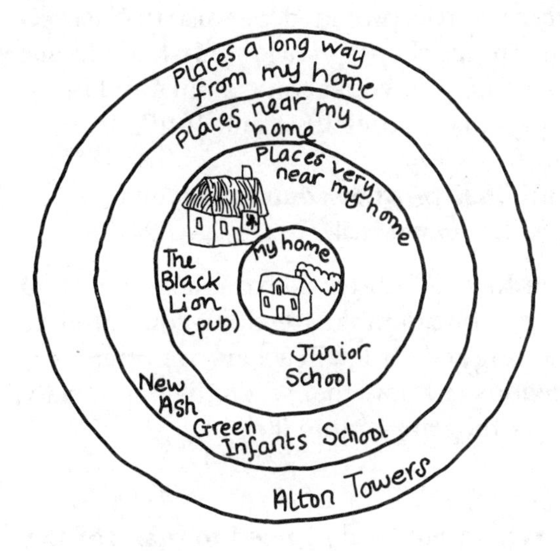

In key stage 2, much work can be done related to the themes of settlement (PoS 8) and the economic activities which occur in settlements (and the theme of economic activities in the Revised Welsh Order).

Children living in the suburb of a town need to have a knowledge and understanding of the visible and invisible patterns which link their locality to the whole town or city (visible patterns such as roads; invisible patterns such as workplace patterns):

- Does their locality provide all the services they use – the library, the fire station, the police station, the hospital? If not, where are these features located?
- What transport patterns link their locality to the rest of the town or city and its region, or, indeed, to the UK?

Figure 10.8

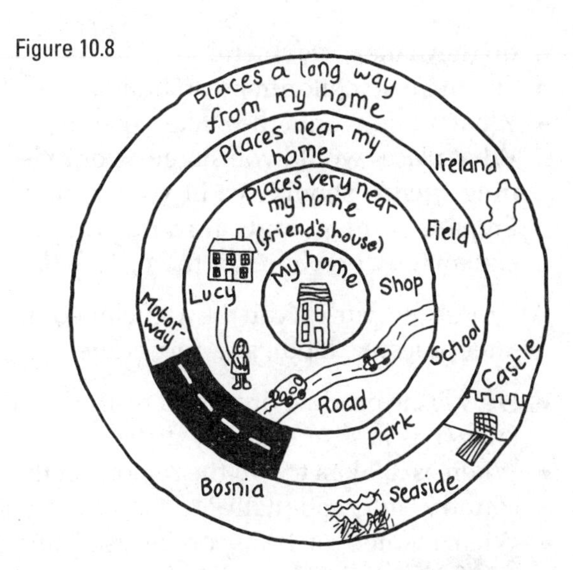

Similarly, children in rural schools need to have a clearer knowledge and understanding of how their settlement and the activities within it are part of a wider pattern. They need to know how it links with nearby villages and towns to understand that their village or hamlet is interdependent with its surroundings:

- Where do they go to the swimming pool, sports or leisure centre?
- Where does the weekly shop take place?
- Where is their dentist?

Starting with children's interests and concerns when situating localities within their wider area is usually the most motivating and successful way to teach:

- Where do parents and parents' friends work, if it is not in the locality?
- Who has moved into the locality recently and where do they come from?
- Where do children's relations live? Do they ever visit them? How (what routes)

do they take to get there?
- To where do children go on holiday?
- What routes would they take to get there? What places would you suggest your visiting friends should see in your region? Prepare a map and a route and an explanatory brochure for them.

All these enquiries can be developed in greater or lesser depth, as appropriate:

- How is the product produced by the local farm/business or industry distributed?
- Where is it taken to – in the region? in the country? internationally?
- What method of transport is used and why? (good links with economic and industrial understanding here) Which routes are used, and why?
- How are goods brought to the local supermarket transported?
- Where do they come from?

It may also be appropriate to emphasise physical geography links with the wider area, depending on the locality. If it is situated on or near a river, then the course of the river, source to mouth, and the river basin area is an obvious way to teach about the wider context of the locality. Maybe your settlement is located within a distinctive geographical area – a natural range of hills, a moorland area, a national park, so photographs, rock samples, maps and the children's own experience can help them begin to understand this.

With contrasting areas overseas, similar opportunities should be taken to situate these places within their wider context. To a certain degree, the extent to which teachers can do this will depend on the teachers' own knowledge and the resources they are using. Usually, it will be easier to develop the concept of the locality within its country and its international setting, rather than within its nearer regional setting. This is not always the case, however. Where teachers have extensive personal knowledge of a country and a place within it, then they will probably have a good knowledge of its regional links. Some resources, such as the Oxfam Cairo pack which looks at four localities within the huge city of Cairo, provide sufficient data to build up a better than average concept of the whole city and its wider context.

The kinds of focused question children can ask when trying to learn more about how an overseas locality links to its wider context will largely depend on those stimulated by the resources provided. See also the last section in this chapter 'Place study knowledge' (page 151) for wider context links and ideas, especially Figures 10.16 and 10.17.

And finally – your contrasting locality queries answered!

The Revised Order is extremely short and gives considerable freedom compared to the original Order. Teachers are often very anxious to know that when they plan place work they are getting it right.

Key stage 1

1 Which *places* do I need to plan for key stage 1?
You should treat the classroom, the school buildings, the school grounds *and* the area around the school as your local area. You need to study one contrasting locality in or beyond the UK.

2 Do I need to plan work in them all for each year?
It is likely but not essential to use all of them in all years, except for the contrasting locality. I would suggest you'll study this once and probably in Year 2.

3 What if school constraints (lack of parents to assist the teacher in supervising

children off-site, problematic children, etc.) don't permit me to go beyond the school gates until Year 2?
This is a shame, but manage it by building in continuity and progression by going outwards from the classroom to the local area from Reception to Year 2. You really must use the school grounds as much as possible in all years. Don't forget that, as we said earlier in this chapter, you can often do fieldwork about the area beyond the school by using the view to the area beyond the school boundary, from inside it.

4 I've wanted to go into the local area in Year 1, but it's just not possible. How can the pupils use their own area in Year 2 to compare similarities and differences with around our school and a contrasting locality?
You can at the very least use a whole range of local snapshots in Year 1 to focus the whole class on common features. You can locate these photographs on a map or model which you and the children have made.

5 What about a contrasting locality in or beyond the UK? Published resources are inadequate and infants only learn by first-hand experience. Piaget said so!

a) Piaget got it wrong for geography – we're proving that daily.
b) Yes, published resources are still a problem in contrasting UK locality.
c) However, there is a huge range of economically developing locality packs available which good infant practitioners use very ably all over the UK to support active, motivating learning about a distant locality. Twinning with another KS1 school will be dealt with later in this chapter under strategies for dealing with distance places.

6 But these are mainly published for key stage 2.
Did you ever meet a key stage 2 teacher who used published materials with no modification or adaptation? The pictures of people and their place in these packs can be used as the principle key stage 1 stimulus.

7 If we want to do a contrasting UK locality and one beyond the UK, can we?
Yes! Wonderful! The one distant place is the statutory minimum – there is nothing to stop you going beyond this, as long as you are already covering all the other aspects of the PoS.

8 We find it difficult to make the distant locality fit into our topics.
Then don't make it fit. Treat it as a half-term topic with a distant place focus.

9 How much time can we give to a contrasting locality place focus?
To develop appropriate depth and begin here/there comparisons, you need to give at least half a term's hours to it, i.e. a minimum of seven hours very specific geography learning objective work, if you operate on a one hour a week timetable. This is a notional time, as, of course, you may block your time into a few weeks.

10 We visit a 'distant' but nearby UK locality. Isn't this the best way of doing the distant locality?
Not necessarily. It can be an excellent strategy – but have you got a range of snapshots of the features you see at the very least – young children need the visual images to prepare or for follow-up after the brief but intense 'visit' experience. Ideally, you need to collect postcards of the place if they exist, along with artefacts, a large-scale aerial view and a large-scale map.

Key stage 2

1 So we have only to do three places, now – the local area, which has got bigger than it was for infants, a contrasting UK local-

ity and an economically developing one. That's the statutory minimum for *localities*, i.e. a small focused area like a village or a suburb of a town or city. You also need to keep situating these localities in their wider context and address all the other contexts mentioned.

2 What do you mean, other contexts?
Paragraph 6 in the programme of studies states very clearly that the UK and the European Union must be specific contexts, as well as a range of contexts from regional to national.

3 Does that mean I can still do some work on my region, as that went really well?
Yes, if you want to. Link it in with PoS 1d – how your local area nests within it, patterns made by links with the rest of the region – rivers, roads, etc.

4 I thought Europe had got the axe...
No, certainly not! It doesn't matter *how* you tackle Europe – through a locality if you've got a good resource pack and/or links with a school, say through learning a modern foreign language like French, or twinning, through looking at a whole country, a range of countries, rivers in Europe: the variations are endless...and with any age of pupils.

5 The UK – does that mean a unit of work on it – could be pretty boring stuff, like the regional geography I had to do at school?
No, not necessarily – it could be a continuing place context in every year, for example 'How do we get to various places in the UK?' in one year, 'What's the weather like in the UK?' in another year, and 'What do we know about settlements in the UK, if we start from the atlas and our own knowledge?' in another year. Every time you are teaching about a locality, region or theme in the UK, don't forget to make sure the children find its relative position on the map of the UK.

6 Could my Year 6 do projects on, say, a European river or rivers, a country in Europe?
Why not? This could now fill the criteria. *But* they must have a geographical framework of key questions, preferably ones they have decided on, to pursue to give them some criteria to work to, not old-style projects which consisted of the flags of European countries,their national costumes, food and stereotypical obsolete customs...They should certainly not just be copying information from books.

7 You keep talking about bits of *themes in places* – rivers in Europe, etc. I thought we needed to look at three *places and four themes*?
No. We need to look at themes in places which is the approach most advisory consultants advise, or places through themes. Read paragraph 6 carefully – that's its crucial message.

8 Should we do more than one theme in a place?
It wouldn't be an effective place study if you didn't deal with some aspects of weather and settlement at the very least. It's up to you – put as much or as little of each theme in each place as you like. Combining aspects of them is the key to integrating geography and developing continuity and progression. Consider the important issues of depth and time.

9 How much time should we give to a contrasting locality study?
A term is a realistic amount of time, especially for upper juniors to achieve the depth. Half a term is the absolute minimum.

10 We go on a residential field trip to develop contrasting UK locality work, but not all the year group wish to come. How do we give entitlement and what about equal opportunities?

a) You have various strategies. None is perfect – you can only do your best. For the majority, whole curriculum benefits must outweigh the geographical and other limitations for the few who are unable to go.
b) Use a contrasting UK locality resource pack while the trip is on, or set different work, not geography, and use the contrasting locality resource pack at a different time, maybe as the other group does follow-up.
c) Use collected resources from the visit – video, photos, postcards, OS maps. Use other pupils as resources to do a less extensive but valued amount of work for those who didn't go.
d) Are you able to arrange a mini-visit to a nearer distant locality for those who don't go with the main trip? If so, use this strategy. Pupils can do their own parallel enquiry into this nearer contrasting locality, with follow-up. Both groups can present to each other.
e) If this is the *second* time you've given access to contrasting locality work in the UK in the key stage, stop worrying about it. Do some other good relevant geography with the group who can't go, and make sure they present it to others. Or do more of another subject area, as long as a reasonable time has been given to geography in this year already.

11 I really don't feel we have adequate resources to do an economically developing locality properly. None of us have been to any of these places, and our library resources are quite inadequate and do not address localities.
There's a tough answer to this one, I'm afraid. Resources are abundant *and* cheap at the price. Lists of these resources are easily available from charity publishers, commercial publishers and probably from your LEA. The resources are generally adequate to enable good in-depth learning to take place. There's no hiding place, if you're not addressing this one.

12 We're a big school with several classes in a year group. Do we all have to study the same localities?
Why should you? It is the principles and process behind the locality study that count. The world is a huge place – you can choose from all the places the resources to address. If everyone does the same place, that's got advantages, as you can share the same ideas, tasks, etc., especially if you're not too confident with it. If you choose different places, it caters for teachers' preferences; and maybe classes could present their localities to each other and so learn more. The key is adequate resources and the principles of learning about places, not the name of the place.

Issues relating to distant place work

The following issues relate to all distant place work in the UK and overseas.

The 'zoom lens' trap The danger of this issue has already been mentioned at the beginning of this chapter. For a strategy to help overcome it, refer to the next section.

'Cook's tour' syndrome Conversely, glib treatment of distant places will, one hopes, be discouraged by the Revised Order. 'Songs from around the world' in infant cross-curricular work do not fulfil the place requirements of the Orders for juniors any more than does a list of basic information on climate, population and famous monuments in a particular town or country.

'Infants can't understand work about distant places.' A few infant colleagues feel strongly that this is still the case. Several arguments counteract this narrow view.

1 Infants are presented with distant-place images constantly via the media and

through books. Often, the only people who will help them to understand these complex and confusing images will be their teachers, who will do so by careful planning and structuring of distant place work.

2 Many infants travel abroad: they have physically had experience of distant places in a holiday context. The teacher can capitalise on this and widen the limited mental picture which often results. Parents do not always do this.
3 Many of our children in multi-cultural areas have a heritage of distant place knowledge and culture. This may be firsthand, or transmitted as their cultural heritage by their families.
4 Attitude formation about other peoples, places and cultures takes place informally from a very young age. To help children develop informed and balanced attitudes we need to discuss other places, peoples and cultures from the youngest age possible in school. This helps to counteract their received prejudice.

Stereotypical images It is important to avoid conveying or reinforcing stereotypical images. The following activity will illustrate the problem.

Task Which ten objects would you send in a box to the distant locality you are studying to show what *your* local area is like? The activity could have two different contexts in the classroom:

1 Developing links with a twin school in a contrasting locality in the UK or abroad
2 When studying a distant locality as a hypothetical exercise – what might we send them and what might we receive from them?

This task could be done either at the beginning or at the end of the unit of work. It can be done with children of any age or with teachers on in-service training. The outcomes will vary according to the age, experience and local knowledge of the participants.

Our selected objects are:
1. a postcard of an oast house
2. a cluster of hops
3. an aerial photograph showing our village
4. an estate agent's leaflet
5. a piece of sandstone
6. a sample of clay soil
7. a local newspaper
8. an Ordnance Survey 1:50,000 map
9. a photograph of our school
10. a tape of traffic noise, bird sound, our language.

There follows an example of the outcome of this task when done with Year 6 children, once a class consensus was reached from initial group discussion.

Anyone receiving such a box of objects is likely to make a stereotypical image of our place. Because you have picked out things which illustrate your area best, you have transferred a stereotypical image. Your country will be viewed as a place where the agriculture is hop farming, or where the natural vegetation is hops growing in clay soil. The other items could similarly foster generalisations. Ten objects sent from North Wales would convey a different image from these from Kent. It is easy to understand where the 'Eskimos live in igloos' stereotype came from. Similarly, if we receive a set of ten photographs – oblique ones and ordinary ones – in a materials pack about Glasgow, which we are studying as a contrasting locality, we must be careful to ensure that the children do not assume that the whole of Scotland or northern UK is the same.

It is crucial for us to realise that when teaching about distant places we must raise awareness of stereotyping in children's learning and break it down, not reinforce it by accident.

Resources Traditionally, because primary schools paid scant attention to distant places, either within the UK or overseas, there was a dearth of appropriate resources. Moreover, the requirement to study localities meant that where primary geography textbooks existed from the 1980s, these are now inappropriate to the scale of locality study required. This is also true of many of the newer texts, which still do not address locality studies effectively because the book format makes the task problematic. Some texts have small sections on contrasting localities, which are also to be found in geography educational TV programmes – demonstrating that fact that the resources for distant locality work need to be as multi-media as possible.

The aid agencies and the Geographical Association have been the principle resource producers for distant localities beyond the UK. They have marketed a vast range of photopacks with accompanying teachers' notes, some with accompanying videos. Educational television has provided the next most useful range of resources. CD-ROM resources such as 'Distant Places' (see Chapter 13) are now coming on to the market and will provide additional resources and learning strategies for pupils.

Resources are improving, but we can reassess what we already have available in our classrooms – postcards, pictures, artefacts from distant places – and look forward to educational broadcasters and publishers continuing to produce appropriate aids. We can also collect postcards and other materials to help ourselves.

With so many contrasting locality resources now available, it is worth spending some time evaluating them, if at all possible, by borrowing copies from other schools where it is not possible to obtain inspection copies. (See Figure 10.9.)

Often the books on geography shelves are

Figure 10.9

Evaluating distant place resources

What are the elements which make up a good quality 'contrasting locality' resource?

- Are there sufficient colour photographic resources for pupils to gain a reasonable image of the physical and human environment of the place?
- Does a video resource also do this?
- Where a video resource exists, does it complement the photos or are the photos video stills?
- Does the resource situate the place within its wider location:
  - town?
  - regional?
  - country?
  - continental?
- Does the resource contain a range of clear maps and plans?
- Does the resource have teachers' notes?
- Do these suggest pupils' activities? Are differentiated activities suggested?
- Are sufficient basic facts about the place given, such as population number and make-up, climate and climate data, capital city and its population, to inform a teacher who has no particular knowledge about the place?
- Are details given about real people, for example a family to whom your pupils can relate?

NB An excellent source of statistics, if those provided are inadequate, is the Dorling Kindersley *World Reference Atlas*, an A-Z fact file and detailed mapping of all the nations of the world (ISBN 07513 01302, £29.95), also on CD-ROM.

CD-ROMs of the atlas and encyclopaedia type will increasingly provide such data.

If you have decided that the basic quality of the pack is sufficient, what additional commercially available resources would you want to acquire for yourself?

- Would the following help?
  - travel brochures
  - Ordnance Survey maps – multiple copies or the foreign country's equivalent
  - TV programme clips, for example from holiday, cookery programmes
  - artefacts – rocks, wool, materials, fruits, etc.
  - commercial postcards
  - tourist literature from the country's tourist office
- What do-it-yourself additional resources could you collect?

The school library should also be audited for its contrasting locality resources.

more of a historical record of distant places decades ago than a resource for learning about localities, countries and continents near the start of the twenty-first century. There is a real need for geography coordinators to check the publication date of books on distant places. If books are more than 15 years old, then the place image as well as the date, text and photo stimulus being presented to children is very inaccurate. Geography is an expensive subject with which to keep up to date! There is a need for drip-feed replacement in the geography parts of libraries, given the resource problems which many schools face.

Bias in resources All the resources we use contain a particular perspective or bias in their text. Travel brochures, often a wonderful source of climate statistics and pictures for a locality, present biased information because they concentrate only on the leisure and exotic nature of a locality. Biased information reinforces stereotypical images, for example, that indigenous people always and only do unskilled work such as waiting or room cleaning in hotels.

Some colleagues prefer not to use aid agency materials for the same reason. They suspect that images and facts have been selected to present the particular image that the agency wants. This attitude cuts off an important source of materials. It is better to be aware that such material could be biased and act accordingly. Examine the materials closely for bias. Ask children questions to bring out any bias; make sure that they are fully aware of that bias.

The photographs in the excellent Action Aid pack 'A Village in India' have a gender bias – most of the photographs show women (see Chapter 13). You need to read the background materials with the pupils and consider why this might be. Whatever their age might be, the majority of children are capable of opening out discussions on values.

Having explored the issues and concerns relating to distant place work, let's now consider the methods of making learning about them practical, motivating and meaningful.

Strategies for teaching and learning about distant places

Key questions The best way of approaching distant place work is to use the key question approach as advocated in Chapter 1, and refined in Chapter 5.

The central questions in Figure 10.4 can be applied to physical or human geography content or to the interaction between both areas of content as shown in the chart. For example, from materials provided about a place in Dominica, it should be possible to answer the question 'What is it like to live here?' 'What kind of farming do the people do here?' could be one of the simple questions young juniors might answer.

You or your pupils can choose the broad key questions or refine them according to your/their experience and their age. The older the children, the greater the number and the more complex the nature of the questions to be asked. Figure 10.10 lists some suitable questions: you will want to select the most relevant and add others, as the list is not exhaustive.

Working towards answers to these questions will deal with the following PoS numbers and paragraphs in a very direct way:

- KS1 and 2 PoS 5a the main physical and human features, and KS2 the environmental issues of places
- KS1 and 2 PoS 5b how localities may be similar and how they may differ

Figure 10.10

Key questions for distant places	
General questions Where is this place?	**Specific questions** Which country, continent, part of the world is it in?
What is this place like? KS1 PoS 5a KS2 PoS 5a	What do people wear? What animals live there? What does it look like? What is the natural and built environment like? What plants grow here naturally? Who lives here? Where do the people live? What do people eat? What kind of homes do people live in? What do people do here? (work/leisure) What is the land being used for? (land use)
Why is it like this? KS1 PoS 1d KS2 PoS 5b	Why do people wear certain clothes? Why do people live here? Why do they live where they live? How have people made use of their environment? In what ways are people's activities and way of life influenced by what we have found out about it? Why do people come here?
How is it changing? KS2 PoS 5d	How have people changed this place? How are they changing it? Are there new buildings or schemes planned? How have new projects altered the landscape or people's lives?
How is it connected to other places? KS1 PoS 1c KS2 PoS 5e	How do people move about there? How do people get to other nearby towns or villages? To where do people send the goods they grow/produce? Do people go to work in nearby places?
How is it similar to/different from my own home village, town, settlement? KS1 PoS 5b KS2 PoS 5b	Is the weather the same as in ...? Is the landscape the same as in ...? Is the farming the same as in ...? Do the people travel mainly by car?
What is it like to be in...?	What do you feel about this place? Why do you think this? Do you think those living there think the changes are a good thing? Why?

- KS1 PoS 5c about the effects of weather on people and their surroundings
- KS2 PoS 5c how the features of places influence the type and location of human activity within them
- KS1 PoS 5d land and building use
- KS1 PoS 5d changes in the locality
- KS2 PoS 5c the wider situation and context of the place.

Depending on the locality chosen and its characteristics and the resources available to teach about it, much thematic work can also be developed through these questions. Theme work should be developed through the context of place work. What is the place like? If your locality has a river which affects it, then this is an obvious context with which to link PoS 7a and/or 7b. Describing and explaining the weather in the locality will evidently develop the weather theme in key stage 2. In key stage 2, the settlement theme is an obvious one to develop here. Environmental change could also be an aspect of the programme of study to work on here in both key stages (KS1 PoS 6 and KS2 PoS 10).

The concept of similarity and difference is crucial to the study of distant places. The key question – 'How is this (distant) place the same as or different from our own area?' – should be implicit or explicit in all our distant place work.

Children working at key stage 1 will concentrate on their locality but will:

- Begin to discover about distant places
- Be able to describe distant places in terms of where they are and what their basic features are
- Be able to deal with mainly implicit, but sometimes explicit, comparisons between the local area and the contrasting locality being considered.

In key stage 2, children will continue to study their locality but will:

- Investigate more distant places
- Begin to analyse – 'How?' and 'Why?' as well as 'Where?' and 'What?'
- Compare and contrast their local area with the distant places being studied to assess similarities and differences between them – explicit comparison
- Begin to recognise, describe and explain processes which change these places.

It is good practice to start with similarities and then proceed to differences. This coincides with the philosophy of Development Education (see Glossary). We all, wherever we live on the globe, have the following common basic needs:

- Shelter
- Water
- Energy sources
- Food.

We may obtain these needs differently, according to the geography and history of our country and culture. We need to emphasise similarity of needs, rather than stress the exotic or curious. Nevertheless, it is ultimately the differences which make geography fascinating. We all 'have' climate and scenery in the UK, but if that climate and scenery were identical to that of the Mediterranean, it is doubtful whether some of us would travel to Turkey or Majorca for a holiday!

A 'here and there' chart, an example of which is included as Figure 10.11, is a useful device for children to demonstrate their learning about distant places and for teachers to assess it. Variations on writing in the chart could include pupils' drawings, cutout pictures, photographs or postcards, and graphical representation according to age, special needs and your differentiation in task according to the children's ability. Note form or whole sentences can be used according to the type of English skills being practised.

Dealing with stereotypes The 'ten objects in a box' activity described earlier raised the issue of stereotyping. We need to begin distant place work with our children with an awareness of what images or stereotypes they have of a country, for example Spain, part of a country, for example Somerset, or of a locality, for example a part of Calcutta. We need also to be aware of what images *we* have of those places, even before we establish our pupils' starting point, so that we can explore attitudes and values freely.

The following approach has been used by the authors:

- With teachers on INSET
- With children in classrooms
- With children to collect images of parents, relatives and other adults.

The approach can apply to any place, anywhere in the world. The more obscure the place, the less a concept of it, inaccurate or otherwise, will be held. The particular example here relates to introducing the Action Aid pack *Chembakolli – A Village in India* to teachers. They were asked to write down in groups of five what came to their minds when they were asked: 'What is it like to live in a village in India?' (See Figure 10.12.)

From the resulting brainstorms, you can see that the images are predominantly negative. Children in areas with little cultural contact with immigrants from the Indian subcontinent have similar images in a more simplified vocabulary.

Where do we get such images from?

- The media:
  - television news (especially at times of catastrophe), documentaries, newspapers
- People:
  - acquaintances who have visited the area
  - acquaintances who have lived there
  - received images from people who have not visited or lived in the area
  - peer group members.

Images of a contrasting locality of the UK could be similarly out of date, inaccurate or stereotypical, for example of South Wales: coal mining; pollution; coal dust; black; spoil heaps; small cramped houses; poverty; steel works; Welsh accents.

Having assessed our own and pupils' starting images of a distant place, we can then ask 'What is it really like to live in a village in India?' or 'What is it really like to live in the Scilly Isles?' By using key questions and up-to-date resources, we can hope both to provide children with a knowledge base from which they can reach a reasonably accurate picture of a distant locality and to shift negative, or even over-positive attitudes, towards an informed view. A useful self-assessment task is shown below.

- Ask children to record their image before starting work on the distant place. Maybe they know nothing because they have not heard of the place. They should also record the outcome of their personal brainstorming.
- Ask children to summarise their learning at the end of the unit of work now that they have some knowledge and understanding.

A format like Figure 10.13 can also be used for teacher assessment and as evaluation for you of the place element of the unit of work you are working through.

Twinning possibilities Twinning with a school in another place is often an effective way of promoting distant places work. Consider establishing a link with a school:

- in a contrasting UK locality
- in a locality in an EU country
- in a locality in a developing country.

You may like to consider the following points about twinning procedure.

- It is best if twinning is a whole-school decision, even though only one year group may pursue it; otherwise the twinning may collapse when the teacher leaves or children go to secondary school.
- If you are nervous about possible language difficulties, remember that in Europe many adults speak good English. Many children in the EU begin to learn English at primary school. Many developing countries in the Commonwealth still use English as their language of instruction.
- Links with developing countries are often made via aid agencies like Action

Figure 10.11

	Here – a Yorkshire town	There – a village on a Caribbean island
Climate and Weather	Cold or cool in winter. Warm in summer. Rain all the year, but most in the winter.	Tropical climate – rainy season July – October, dry season December – June. Hurricanes may occur.
Landscape	Flat, with wide river meandering across plain. Hills rise up in the distance.	Hilly, with a volcano called Mount Pelee. Dormant volcano.
Vegetation	Grass on hills, near river. Trees divide some fields.	Tropical rainforest to north, fields and plantations everywhere else.
Farming	Sheep on hills, crops away from river. Barley.	Growing bananas, pineapples on plantations. Some sugar cane grown.
Industry	Glassworks, paint factory. Lorry depot. Lots of people out of work.	Very little. Some rum production. Lots of people out of work.
Travel	By car, few buses. Good rail links to York and London.	Bus or shared taxi. Transport cheap. You don't have to wait long. No railway.
Any environmental concerns, projects	They want to build a by-pass to take lorries out of the town.	Local people want proposed hotel built to provide more work; mayor does not want environment spoiled for tourists from USA, Europe and Venezuela.

This format can be given as guidance for children to discuss similarities and differences first, then to record individually later. Will they always need to record?

Infants can try the charts without the categories – just 'here' and 'there' columns.

Who decides the heading: pupil or teacher? How many headings?

The younger the child, the fewer the headings?

What form should the recording take? Note form? Sentences?

Could a combination of writing, drawings, stuck-on pictures, sketch maps or/and diagrams be appropriate?

What are the most appropriate methods of comparing here/there for younger pupils or special needs pupils: drawings, writing, etc?

Figure 10.12

Outcome of a brainstorm: 'What is it like to live in a village in India?'

Images of India: positive? negative? neutral? fact? fiction? from the media? received images?

poor hygiene, outdoor cooking, insects, undernourished animals, primitive transport, poor schooling, religious or culture-based clothing

begging, temples, working in fields, Mother Teresa, sacred cow, rice, tea, early death, dhoti, burning widows, silver jewellery, overpopulation, smelly, hot, wet

poor sanitation, no electricity, dust, dirt, barefoot, poverty, disease, hot, humid, communal living, extended family

bikes, shanty towns, cows, sparse vegetation, lack of equipment, poor health, heat, no amenities (shops, sewage, electricity, water), dust, flies

Figure 10.13

Before What I think about ....before I learn about it	**After** What I know and think about .... now

Aid and church organisations, as then our values cannot directly damage the less materialistic values of children in developing countries.

- Schools in developing countries may have smaller budgets than British schools: they may have neither the technology nor the cash to respond with photographs, tapes or videos. Letters and drawings are often the most they can manage.

Once you've twinned, consider the range of possibilities in Figure 10.14.

If you are twinning with a school and intend to visit it as well as exchange materials, then it is essential that the contact time at the twin school should be very carefully programmed. The more contrasting the background and locality, the more carefully both sets of staff and children should be prepared. Contrasting backgrounds in the UK can cause culture shock if the meeting is not carefully managed. It is better if preparation work in the form of exchange letters and photographs has occurred as an absolute minimum, so that children have a specific 'twin' to relate to when they meet.

Here are some examples of particularly successful distant place activities achieved through twinning.

1 Two Year 4 classes exchange within their contrasting UK localities.
One school is located inland in a route-centre town with Norman origins, the other 60 miles away in a small coastal resort with Georgian seafront and some deep-sea fishing still taking place.
Children did extensive project work on their own locality, shared it with parents, then sent it to the visiting school. The two schools used exchanged project work as the starting point for their visits, did further preparation on their distant locality, then were guided around their host school's locality with learning objectives relating to geography and history. Teachers and children were tremendously enthusiastic about the outcomes of the visit and the whole project.

2 An English school twins with a school for 7–11 year olds in Utrecht, Holland.
An English Year 5/6 vertically grouped class visited their local town centre, identified types of shops and mapped the layout of the interiors of some of the shops. Materials were sent to Holland

Figure 10.14

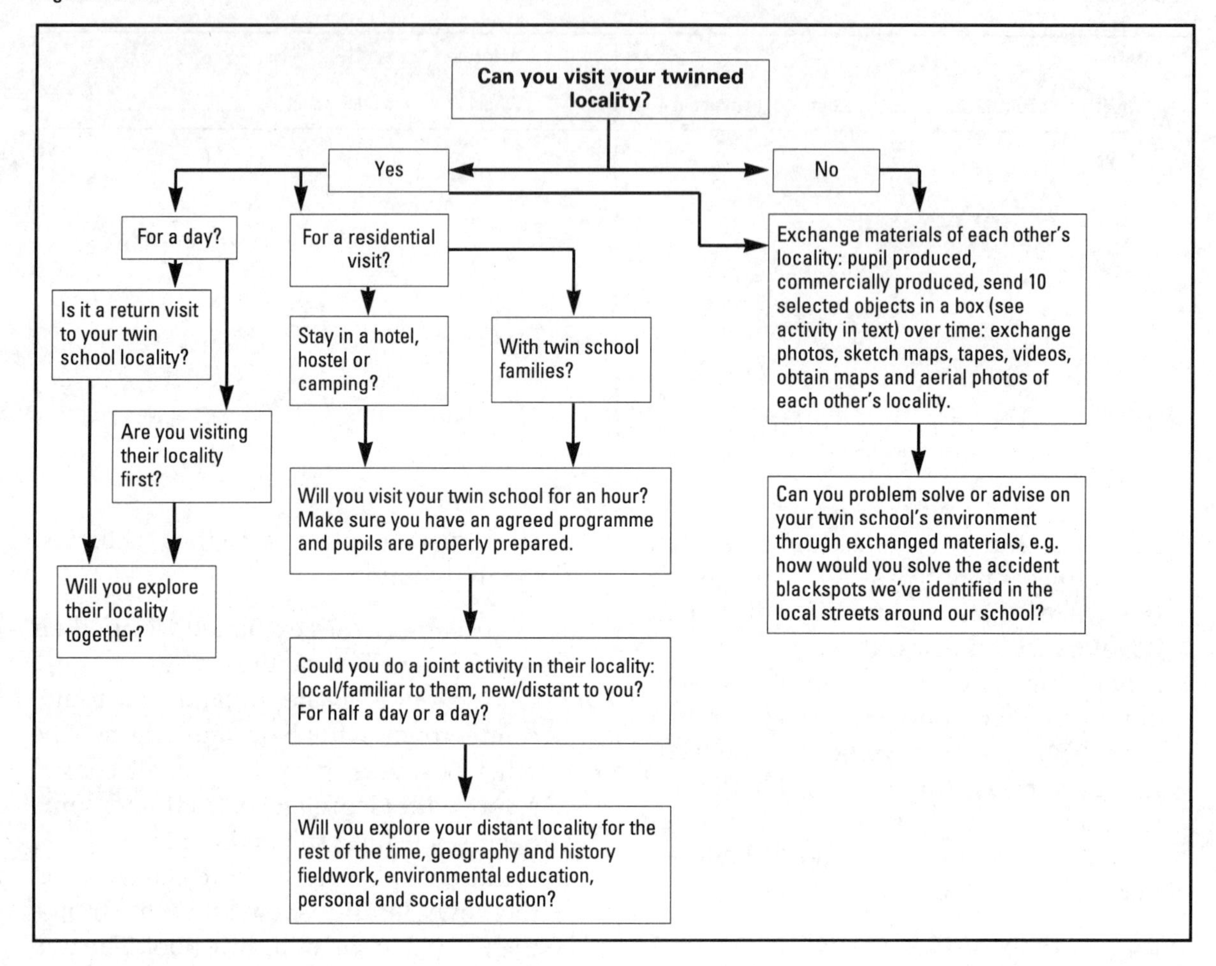

with the questions: 'Do you have similar shops to ours?' 'Are they stocked and laid out in the same way?' The written, graphed and mapped responses changed the English children's ideas: they found great similarity in the type and layout of shops, whereas they had expected differences.

3 Two classes, one in a northern England urban location and one in a south west England rural location, could not visit each other, but exchanged materials with their IT resources and by electronic mail, which speeds up communication. Sketch maps and other graphical materials were sent on behalf of the school by fax, via two parents working from home with whom payment was negotiated.

Collecting and using appropriate resources Michael Storm, a member of the Geography Working Group, has provided the following very helpful list which enables us to focus on the kinds of resources we need to gather.

To carry out locality studies, we need:

- A specific, named location (i.e. not India but Chembakolli, a village in India)

- Named people, preferably families
- A focus upon the lives of children
- Pictures of people, landscapes, buildings and unfamiliar artefacts, crops, etc.
- Maps and plans, from small-scale contextual locating maps (e.g. India, South India) to large-scale maps (e.g. Chembakolli village) and house plans
- The pattern of daily life – 24-hour time lines, etc.
- How income is earned, or subsistence organised;
- Diet, clothing, housing details
- Patterns of movement: work, school, trade
- Shopping, market, trade activities
- Leisure activities, festivities, special occasions
- Connections with the wider world through trade, travel, media
- Changes, recent/imminent, in landscape and life styles: aspirations and problems
- Data on climate, descriptions of weather
- Descriptive or imaginative literature set in the locality or its region.

Some characteristic pupil activities In planning an imaginary journey to the locality (route, time, cost) children can be involved in the activities listed below.

- Making posters to 'advertise' the locality
- Making pictures of the locality as envisaged
- Writing a letter home from an imagined visit to the locality
- Writing to a child who lives in the locality
- Writing a journal/diary entry for an imagined visit
- Writing a story set in the locality
- Making a map of the locality and its setting (using symbols, key, grid, index, scale)
- Making a model of the locality or a building within it
- Making diagrams (time-bar) of the daily rhythm of life
- Making calendar diagrams (pie chart) of the annual rhythm of life
- Making climate graphs for the locality
- Debating a proposed change in the locality
- Writing a letter to an envisaged local paper or enquiry
- Collecting and sorting other material about the country in which the locality is situated, e.g. pictures, stamps, coins, flags, products, etc.

Primary Geography Matters,
(The Geographical Association, 1991)

You probably have many resources which you will need for studying distant places, but will need further specific ones.

For key stage 1 resources see Figure 10.15.

In addition:

Figure 10.15

Resources for key stage 1 for studying distant places

Resource	Notes	Resource	Notes
• Photographs	your own and commercial	• Maps	
• Pictures	e.g. from calendars, posters	• Artefacts,	e.g. pottery, hat, length of material, clothes
• Postcards		• Travel brochures	useful, but biased
• Stories	situated in distant locations	• Rock or fossil samples	
• Information books	about countries, places, environments	• Big book-type atlas	
• Slides	your own and commercial	• Infant atlases	
• Radio/ TV programmes	best recorded for selective use		

NB Your own photos can be enlarged on a colour photocopier

- Children may be able to lend resources from home
- You may add to your own collection when you travel abroad
- Supermarkets and ethnic food stores are wonderful sources of fruit, vegetables, spices and other foods from distant locations
- A visitor or local contact from the distant locality invited into school has long been an excellent resource
- If you have a child in your class who is visiting relations abroad, enlist them and their parents to collect resources for you while they are there. They will probably be delighted to be of help, but remember to negotiate a budget for them. You will need to specify your requirements and decide whether it is feasible for them to collect some or all of them. Refer back to the checklist of optimum ingredients on pages 148–9.

Remember to use your resources not just to develop work on the localities required in the part of the programme of study relating to places, but also to develop awareness of the wider and more general range of places inferred in the themes and in key stage 1 PoS 1c, key stage 2 PoS 1a and 1d.

Use photographs and/or slides to develop the recognition of:

- The *physical* environment in distant places
  - a river photograph
  - a waterfall photograph
  - a volcano photograph
  - a wind photograph (for example a wind pump, a wind surfer)
  - a rain photograph (umbrella)
  - a sun photograph (an umbrella used as a parasol)
- *Human* environment activities
  - a work photograph (for example hand ploughing in a rice field)
  - a communications photograph (for example illustrating the use of bicycles)
  - a settlement photograph (an oblique aerial view of a village or part of a town).

Children talking about these photographs will develop their knowledge and understanding of vocabulary of:

- Aspects of distant places
- Geographical skills, such as direction finding and use of oblique aerial photographs, as many photographs and postcards are a form of oblique aerial photograph.

For key stage 2 a similar list of resources applies, but increasing numbers of the following are needed:

- Atlases
- Information books or 'text-type' books with details on localities
- Information packs on localities
- Larger-scale maps of localities (commercial or sketch maps).

Refer back to Figure 10.9 for the evaluation of new distant place resources.

Fiction and poetry are also important resources which could be listed as school resources for learning about distant places.

Fiction and distant places

We habitually read to our classes or encourage children to read fiction which is set in both time and place. With a little refocusing, we can capitalise on such fiction to develop history and geography work, while still ensuring that the enjoyment of the book for itself is most important.

All the fiction we read to children has some sort of 'place' background; but in some novels, it is especially relevant to heighten the sense of the place in which the novel is set. Fiction can enhance geographical learning, if:

- Children are encouraged to locate the place in the book and its surroundings in an atlas
- Descriptions of physical features or processes are made explicit and reinforced by the teacher, for example a volcano
- Descriptions of human processes and patterns are similarly clarified, for example farming, settlement
- The meaning of specific geographical vocabulary used in the book is checked and/or clarified by the teacher, for example a railway viaduct.

The following passage from *The Cay* by Theodore Taylor (Puffin), set on a tiny island in the Caribbean, illustrates the usefulness of fiction for geographical learning.

A description of erosion by water:

> *"He described the hole to me. It was about twenty feet in diameter and six to eight feet deep. The bottom was sandy, but mostly free of coral so that my hook would not snag. He said there was a 'mos', natural opening to the sea, so that the fish could swim in and out of this coral-walled pool.*
>
> *"He took may hand to have me feel all around the edges of the hole. The coral had been smoothed over by centuries of sea wash. Timothy said the sand in the sea water acted like a grindstone on the sharp edges of the coral. It was not completely smooth but there were no jagged edges sticking out.*

It will be necessary to clarify geographical vocabulary: the 'cay' as in Florida Keys and 'volcano'.

> *"We often talked about the cay and what was on it. Timothy had not thought much about it. He took it for granted that the cay was always there, but I told him about geography, and how maybe a volcano could have caused the Devil's Mouth. He'd listen in fascination, almost speechless.*
>
> *"We talked about how the little coral animals might have been building the foundations of the cay for thousand of years. I said, 'Then sand began to gather on it, and after more years, it was finally an island.'"*

The passage also describes the process of coastal deposition, which could be contrasted with the process of erosion in the previous passage. Erosion and deposition occurs in KS2 programme of study 7b, related to rivers but may also relate to 5a – physical features in localities, which may be coastal. Poetry can also be a very useful medium for teaching about distant places, as the following poem illustrates. Such a poem could be a small starting point for a great deal of geography about a Jamaican locality. Clearly maps, photographs and further details will be needed, but the essence of the effects of weather and climate on the island's human geography is here.

> *"Year in, year out, the land looks up and waits.*
> *Year in, year out, the land is battered by the slanting rain*
> *Which softens the brain, the earth, roots the sugar cane,*
> *Washes away the top soil, breeds angry mosquitoes,*
> *The land is flattened by hurricanes, like pneumatic drills,*
> *Which uproot ancient trees, smash houses,*
> *Splinter the sleepers on railway tracks.*
> *Whiten the corners of hungry mouths,*
> *And drown the population, given half the chance."*

From a poem by a Jamaican poet. Preface to *Hurricane* by Andrew Salkey (Puffin)

Place study knowledge

Revised Geography in the National Curriculum is quite specific in the programmes of study about the particular locational knowledge which children are expected to acquire by the end of key stages 1 and 2.

This knowledge is expressed in the form of maps at the end of the key stage 2 programme of study with particular physical features and towns which children have to be able to identify:

- UK – Map A
- Europe – Map B
- World – Map C (see Chapter 2).

Each programme of study also has a subsection under PoS 3 (Geographical skills) which highlights particular place knowledge:

- KS1 PoS 3c: locating and naming on a map the constituent countries of the United Kingdom, marking on a map approximately where they live
- KS2 PoS 3d: identifying the points of reference specified on maps A, B and C.

The authors have tried to emphasise throughout this book that the approach to the acquisition of this place study knowledge should be developed through atlas and globe work in the context of human and physical environments, not as 'one-off' rote-learning. Place and theme should always be used to widen locational knowledge and spatial awareness.

The key question 'Where is the place?' and its related concept of location refers, of course, to both place study knowledge and the 'zoom lens out to wide-angle lens' focus issues described earlier in this chapter. It is important to know where places are, but part of that knowing should be understanding *where* a place is *relative* to another place or the region, country, continent or global sector in which it is situated. Learning about place is a process. Much of the formal locational knowledge required by the maps will be acquired when localities are studied if teachers adopt the approach advised here, locating each locality in its wider sphere and encouraging children to see how it is connected to other places, for example capital cities and other countries.

Children should also be aware of the process required to find out where a place is, including how to look up where that place is in an atlas via the contents list, index, coordinates and grid references.

The pupil activity suggestions in Figures

Figure 10.16

Suggested questions to develop spatial and locational understanding	
Direct experience	**Indirect experience**
If the pupil has visited a place use these types of questions:	If the place being studied cannot be visited, adapt the questions in this way:
UK	*UK*
1 How did you travel there?	1 How could you travel there?
2 What counties did you travel through?	2 Which route could you take from here?
3 Which county did you stay in?	3 In which county is the locality?
4 If you stayed on the coast which sea did you swim in?	4 Which sea or lake is nearest to the locality?
5 Which other countries were nearest to you and which seas surround the UK's coastline?	5 Which other countries and seas border England/Wales/Scotland, and which seas surround its coastline?
Abroad	*Abroad*
1 How did you travel there?	1 How could you travel there?
2 Did you travel through or fly over any other countries?	2 Which other countries would you be likely to fly over or travel through?
3 Which seas did you cross?	3 Which seas might you cross?
4 If you visited an island which seas surrounded it?	4 If an island, which seas surround it?
5 If not, which countries were next to/bordered your holiday country and which seas washed its coasts?	5 Which countries and seas border your distant locality?

Figure 10.17

Suggestions for pupil activity

1 Pupil shades or marks the country under discussion on a world map.

2 Pupil uses or leaves a map of Europe according to location of place being studied.

3 Pupil reinforces their own location by marking on a map of UK.

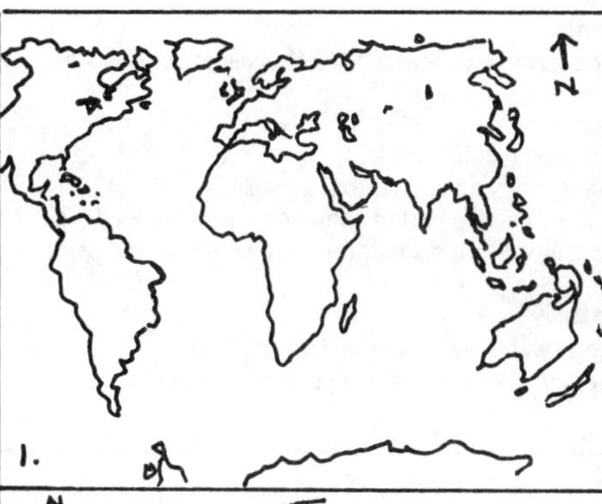

N

To U.S.A

Cuba

Puerto Rico

To Mexico

Dominica

CARIBBEAN

Martinique

SEA

Barbados

Trinidad

Caracas

Venezuela

4.

Where is Martinique?

Martinque is one of the Windward Islands. It is in the Caribbean Sea, in the eastern part. Its east coast is washed by the Atlantic, the west coast by the Caribbean.

You have to travel south west from England to get there. The nearest big country is Venezuela in South America.

Its capital is Fort de France.

5.

4 Ask pupil to draw a sketch of the country the locality is situated in and to mark it on. Teacher or pupil sets detail required.

5 Ask pupil to write a variety of information according to their ability and age.

You might consider directed/undirected activity.

'Fill in the blanks' activity with vocabulary might be provided by teacher or chosen by child.

NB Maps 1, 2 and 3 are already on the worksheet, prepared by the teacher. 4 and 5 are completed by the children.

10.16 and 10.17 could be modified/adapted to help children develop spatial and locational understanding while using atlases in context.

The essence of teaching about places

The points in Figure 10.18 will serve as a check-list for teaching about distant localities, within the UK or beyond.

Figure 10.18

Teaching about places

Why teach about places?
- worthwhile
- relevant
- enables balanced views to evolve
- reduces insularity

Try to ensure the study is
- realistic with real people and places
- linked to the child's own perception of local and distant places

Try to avoid
- bias and giving one-sided viewpoints
- stereotyped images
- token gestures
- 'Cook's tour' syndrome

Try to develop twinning
- locally
- nationally
- wider world
- links can be with individuals, schools, villages or visitors

Remember
- distant may be within the UK
- that a place study can arise from a theme
- a place study may be part of another subject-focused topic
- a place study may be a separate and discrete unit of work

Depth of study
- key stage 1 awareness/general and one locality
- key stage 2 greater understanding/prescription/depth

11

KEY AND CROSS-CURRICULAR SKILLS IN GEOGRAPHY

This book stresses the first-hand experience and 'enquiry process' approach to geography as the one which leads to the most effective learning and teaching. It also supports children's development in oracy, literacy and numeracy referred to as *key skills*.

Primary geography also requires the use of many secondary sources to back up and develop fieldwork and to provide knowledge and understanding of people and places which it is impossible for children to visit.

As geography is the subject which bridges the gap between the arts and sciences, the possibilities for developing and reinforcing pupils' skills are limitless.

The skills of using secondary sources refer in some measure to all subjects across the curriculum. This chapter briefly outlines opportunities for developing them in a geographical context.

Research skills

Pupils will need, by the end of their primary years, to be able to read and extract information for a purpose from a range of written resources, including:

- Tourist leaflets
- Travel brochures
- Newspapers
- Information books
- CD-ROMs
- Encyclopaedias
- Story books with a geographical setting.

They will need to work towards the development of higher reference and reading skills such as:

- Using a contents page
- Using an index
- Extracting factual information from more than one book
- Scan reading
- Skimming text
- Summarising
- Detecting bias or stereotyping in text
- Detecting inaccuracies in text
- Ascertaining when a book was published.

The last is very relevant to geography, as many geographical reference books in class and in school libraries have themselves become historical documents: children need to know how up to date their information is!

Graphicacy skills

Geography uses graphicacy as a source of secondary information more than any other subject except maths.

Pupils will need to understand and extract information from:

- Graphs
  - block graphs
  - bar-line graphs
  - line graphs
- Charts
  - bar charts
  - tally charts
  - frequency charts
  - pie charts
  - flow charts
- Tables of various sets of data
- Maps of the many kinds already discussed.

Media sources

Educational television and radio, films, filmstrips and slides provide an important source of secondary information for pupils. Extracting the most useful information from these sources for geography can be a difficult skill. Pupils need to watch and listen, having been properly briefed on the programme. It may be appropriate for the children to watch with one or more key questions in mind to enable them to focus on a particular concept or area of content. Alternatively a useful technique with older children is to watch some programmes twice, the first time with no brief, but the second time with key questions in mind, or, for the more able children, a requirement to take some notes. This should not, however, be at the expense of watching, and the pause button should be used.

Some examples of key questions:

- What are the homes like there?
- What kind of work are people doing?
- How many different kinds of rock or landscape can you see? Note them down.

Whatever secondary sources have been used, teachers and children should know what the expected outcomes of the learning are. Information should be extracted not just *per se*, but to inform the pupils and to lead on to pupil comment, analysis and prediction, depending upon pupils' age and ability.

Data handling

Children collect information in the form of facts and figures from their fieldwork and secondary source enquiries. Traditionally, this data has been dealt with manually.

Findings are recorded, reflected on, questioned and so on by writing or using the various graphicacy skills. Some work will continue to be done like this but, whenever possible and appropriate, we need to encourage the use of IT as a key tool in both handling and presenting data. The presentation of data through electronic means is a bonus for geography, and helps pupils to spend more time on analysing their results than, say, drawing a block graph or pie chart.

The National Council for Educational Technology's *Focus on IT* states that handling data with computers should encourage the ability to:

- Plan, hypothesise and predict
- Design and carry out investigations
- Interpret results and findings
- Draw inferences
- Communicate ideas and findings
- Make decisions based on conclusions drawn from findings.

This list of aims could equally apply to geographical enquiry. IT provides the tool to handle the data or information generated by a geographical enquiry. Pupils will need to learn how to store, retrieve, correct, process and present information. Database packages like *Grass* and *Our Facts* are exceptionally good for primary age range children. *Our Facts* is a simple 'user friendly' package which can be used by pupils from Year 1 on, as it can be set up with limited fields. It is also useful as an introduction for older pupils and teachers who have no previous experience of a database. *Grass* is a slightly more complicated package for pupils from about Year 4 onwards. It has 16 fields, and can do complicated searches and mathematics. Both these programs are the box-file type of data package. Pupils can also be taught to access public data bases such as Ceefax, Teletext and library systems. Spread sheets can be used especially to plot field work data on and plan environments.

Touch Explorer Plus using the concept keyboard is more recent and is an excellent addition to the software available for juniors and special needs children. Its various files allow the graphic representation of data collected during fieldwork or secondary source research to be represented on a base plan of A3 or A4 size. The plan is divided up by a grid and information stored electronically under the plan at different 'levels'. These 'levels' can be drawn up to the screen by a 'touch' on a section of the plan or picture. This program allows for differentiation in levels of difficulty and/or type of information stored below the plan. For example, in a land use survey on a local high street, the first level of information could be the location of shops, the second could be the type of shop – for example butcher, estate agent, timber yard – and the third level would explain exactly what the shops did – for example estate agent, a service industry selling or renting houses; a timber yard, where you can buy woods, nails, fencing, and so on. The children can write the information to be stored in the different 'levels'. They can write for different audiences – for example classes lower down the school, different children in the same class and maybe adults.

The open-ended possibilities are too numerous to list here, but most aspects of data from physical and human geography or environmental considerations can be used. Publications of IT programs are given in Chapter 13.

Presentation skills

Geographical work lends itself to every method of presentation. The following very varied methods of presentation should prove useful to teachers in the context of special needs children as well as mainstream pupils.

Depending on the special need in question, differing methods of presentation from those used by the majority of children can be used by a child to demonstrate understanding in geography.

Written presentation

- Children can record factual information with further analysis or comment.
- They can integrate factual information into creative writing – a story or a poem in a geographical context, for example, a story in the geographical setting of an Amazon location after a project on the equatorial environment; 'I'm only a pebble on the beach': creative work tracking the historical and geographical progress of a flint pebble from cliff to stack to beach after work on coastal erosion and deposition.

Graphical presentation

Maps and plans are often needed to clarify written work or to serve the purpose of written work.

- IT will speed up many types of graphical presentation.
- Charts, graphs and diagrams can present data which has been collected and then analysed. Flow charts, pie charts, here-and-there comparison charts for places are a few examples (see Chapter 10, Figure 10.8 for here-and-there charts).
- Landscape sketches and general drawings can be made.
- Use base maps as data collection sheets.

Spoken presentation

Spoken presentation can include:

- Straightforward reporting back on experience or evidence
- Tape recording
- Role play – in a geographical context, each child may play the part of a local resident or interested party to simulate a public enquiry surrounding a local issue, such as a bypass or reservoir scheme, to culminate work done on the issue and to demonstrate understanding.

Models

A great variety of models with art or design technology connections can be used to present and enhance geographical work.

Some examples are:

- 3D models of villages, streets, contour models, volcano models, cliff cross-section models
- Working models of water mills, weather instruments.

Photographic presentation

Children's own photographs and postcards can be used to:

- Enhance a class or group display of geographical work
- Enhance their own project books.

They should be used for a purpose – annotated to explain children's understanding, and occasionally for fun, not just stuck in to fill space. One very good use of photographs is to mount them next to a rough sketch map which shows the location at which each one was taken and notes in which direction the camera was facing when it was taken. Pupils can annotate features, too, if appropriate.

On rare occasions, a video taken during fieldwork or in the classroom can be directed by the children.

12

CROSS-CURRICULAR THEMES AND DIMENSIONS IN GEOGRAPHY

'How does effective National Curriculum geography teaching deal with cross-curricular themes and dimensions without entailing further work?' This is a reasonable question which teachers have the right to ask.

To answer it we need to clarify what the themes and dimensions are; to recognise their relationship with geography and to focus on individual themes and plan them into geographical work where appropriate. This does not require a lot of extra work, but a heightening of our own awareness to bring out the explicit nature of the themes with pupils. Sometimes a change of focus within already planned work will be required to achieve this.

The cross-curricular themes

The five themes are:

- Economic and Industrial Understanding (NCC 4)
- Health Education (NCC 5)
- Careers Education (NCC 6)
- Environmental Education (NCC 7)
- Citizenship (NCC 8).

The Whole Curriculum, NCC 3, gives an outline of cross-curricular dimensions, skills and themes for quick reference.

For further, more detailed reference, the National Curriculum Council (NCC) sent one copy of all these Curriculum Guidance themes to all maintained schools between late 1989 and the end of 1990.

In many primary schools it is not unusual to find that these books have sunk without trace. However, to understand the concepts of each one in detail you will need to hunt them down and find the time to read them. Further copies are available from SCAA Publications. As a coordinator, this is a useful activity to build into your action plan (see Chapter 14), as someone in the school should have a deeper knowledge in order to advise others.

If you analyse these NCC guidance books in relation to *primary* geography you will see that the relationship between:

- Environmental Education and geography is very strong
- Economic and Industrial Understanding and geography is very strong
- Citizenship and geography is important
- Health Education and geography is contributory
- Careers Education and geography is contributory.

All the themes have a strong relationship with human geography because it is human action in and upon the environment which gives rise to the themes from the geography stance.

Environmental Education within geography

Environmental Education is of great importance to geography. The geography Order has highlighted part of the human environment interaction with the physical environment content of the subject by pulling out these aspects, including specific environmental themes as part of the Order.

In key stage 1, when investigating the locality of the school and a contrasting locality, one in the UK and one abroad, children should be taught, as in KS1 PoS 6, the thematic study:

- To express views on the attractive and unattractive features of the environment they are studying
- How that environment is changing
- How its quality can be sustained and improved.

In key stage 2, PoS 10, the Environmental Change theme, whilst investigating the locality of the school, and two further contrasting localities (one in the UK and one in a developing country), pupils should similarly be taught:

- How people affect their environment
- How and why people seek to manage and sustain their environment.

Thus, Environmental Education is a statutory part of the Geography curriculum expressed clearly through thematic studies.

However, it is important to look beyond the environmental themes to the other parts of the programmes of study. Environmental Education is not only an integral part of the places being studied: it also exists, either explicitly or implicitly, in the other themes, In key stage 1, for example, KS1 PoS 5 is important. Pupils cannot express their views about environmental quality until they have some understanding of KS1 PoS 5a – what the main physical and human features of their environment are like.

In KS2 PoS 5a, similarly, pupils have to know about the main physical and human features and the environmental issues which give localities their character. In work relating primarily to settlement, pupils need to know how land is used in settlements (PoS 9b) and how they can go on to learn about any particular issues arising from the way land is used.

Environmental Education is defined as education:

- *In* the environment
  - active learning
  - fieldwork
- *About* the environment
  - to gather knowledge and understanding
- *For* the environment
  - in order to preserve, conserve and respect it.

The basic aims of geographical work share these ideas.

At the time of writing, SCAA, RSPB and the Geographical Association are preparing publications on Environmental Education.

Economic and Industrial Understanding within geography

Economic and Industrial Understanding is concerned with:

- The nature of goods and services provided to customers
- How these goods and services are provided
- How some leisure activities harm areas of environmental value
- Economic activity – farming, industry, business and the world of work
- Industrial location – why certain economic activities are located near to a

power source, workforce, raw materials, a transport network, consumer market or other resource
- The impact of industry on the local area
- The effect of economic change on the local area and people
- The economic importance of renewable and non-renewable resources
- The effects of extracting natural resources from the environment
- The impact of new industry on the environment.

Citizenship within geography

Citizenship is concerned with:
- Similarities and differences in communities in different places and also on every scale
  - local area
  - regional
  - contrasting locality
  - countries
  - global
- Similarities and differences in the quality of life
- Decision-making processes
- Issues relating to change in settlements
- Attitudes and values and their effect on change.

Health Education within geography

Health Education is concerned with:

- Hazard perception and safety in different fieldwork environments
- The prevention of pollution in water sources in the UK and in economically developing localities
- How the local physical geography and standards of wealth can affect the lives of people in different places
- The quality of environments and how this can affect the health of communities
- Conserving the Earth for our own health.

Careers Education within geography

Careers Education is concerned with:
- Different types of work people do
- Differences in types of work, opportunities to work and reasons for this
- Environmental problems arising from industry.

Figure 12.1 relates the programme of study elements to the various cross-curricular themes in more detail.

Some examples of activities which develop cross-curricular themes at key stages 1 and 2

Having recognised what the themes are and how they relate to primary geography, the following examples are given to illustrate how the themes can be highlighted in geographical work already being done. Many teachers are resistant to the themes at first glance because they do not feel, except perhaps for Environmental Education, that they are appropriate for young children. This is a limited view, as many of the activities which we are already doing with pupils involve the themes. We just need to recognise this and make it explicit, while, however, ensuring progression, of course. We can do this by extending our open-ended questioning of children's understanding and by slightly altering the focus of part of a topic.

Environmental Education
Key stage 1
- Children can map the worst litter zones in the school grounds and around the school, or
- plan a scheme to improve the school grounds: working with environmental awareness.

Key stage 2
- In an ongoing topic relating to the change in use of a local site, pupils can

Figure 12.1

Relating the geography Order to cross-curricular themes at KS1 and KS2

Programme of study elements	Economic and industrial understanding	Environmental education	Citizenship	Health education	Careers education and guidance
SKILLS					
Map use	△	△	△	△	△
Fieldwork	△	●	△	△	△
PLACES					
Knowledge of places	△	△		△	
Understanding of distinctive features	●	△	△	△	
Similarities/differences	●	△	●	△	
Themes/issues	●	●	●	△	
Changes in places	●	●	△	△	△
PHYSICAL GEOGRAPHY THEMES					
Weather and climate		△		△	
Seas and oceans		△			
Rivers, river basins	△	●	△	△	
Landforms		△			
Animals, plants, soils		●			
HUMAN GEOGRAPHY THEMES					
Settlement	△	●		△	
Communication and movement	△	△			
Economic activities	●	●	●	△	△
ENVIRONMENTAL GEOGRAPHY THEME					
People's effects on the environment	△	●	△		
Managing and sustaining environments	△	●	△		

△ = links ● = strong links

monitor the restoration of the landscape on and around the site over a period of time.

Economic and Industrial Understanding
Key stage 1

Infants frequently set up shops and businesses in the classroom and may visit real ones as fieldwork experience.

- Set up a travel agency with top infants as a way of beginning work on a particular locality beyond the UK, using the globe to locate countries and differentiate between land and sea. Have pupils make mock cheques for fictitious or real costs, depending on their place value concepts, for their journeys or package tours.
- When visiting a dairy farm, make sure that the children understand that farmers need to buy and sell calves or cows and buy feedstuffs in order to produce the milk pupils see.

Key stage 2

- Children preparing for a residential journey for fieldwork can be involved in the costing of those activities, including alternative transport costings against distance. They can compare the group entrance fees for places they will visit and consider the difference in likely maintenance costings for different places, for example which costs more to run – a castle or a National Park visitors' centre?

Local issues

- Pupils examining the issue of alternative by-pass routes, actual or fictitious, can contact the local highways department to discover construction costs per mile over different land types, so that they can make a more informed decision about the costs and benefits of the alternatives in their public enquiry simulation.

Global issues

- Upper juniors studying an economically developing locality in an equatorial or tropical zone need to be made aware of, and consider both sides of, the deforestation problem. Conservation, world climate and soil erosion versus the economic pressures on tropical governments to export hardwood to keep part of their population employed and to trade with other governments are important issues. Bringing in much needed foreign currency resources benefits the population and needs to be considered.
  Here the area of economic needs, as opposed to wants, costs and benefits, environmental and social issues associated with economic activity, is developed.
- Within a topic on conversation is a focus on animal conversation in the Arctic. It could address the issues that some Inuit need to hunt to make a living and that it is appropriate for indigenous people to dress in fur clothing in the sub-zero temperature zones of the world.

Economic and Industrial Understanding is often considered a particularly difficult theme for primary pupils. The above suggestions should disprove that theory. After all, infants able to understand that if there are not enough pairs of scissors in their class for one pair each then they must share are already understanding the concept of scarcity of resources, a concept central to economics. Similarly, juniors can understand that when the money inputs of a car manufacturing industry fall because customers cannot afford to buy cars, then car workers are laid off.

Economic and Industrial Understanding can enhance geographical work because the two lend themselves to the involvement of Adults Other than Teachers (AOTs). The teacher does not and cannot be expected to know the details of other people's work which could enhance their pupils' learning.

Pupils engaged in a geographical enquiry can gain invaluable benefits from either visiting professionals or asking them into school to help with the enquiry. Professionals, however, should *not* just act as one-off deliverers of information. As part of their enquiry, pupils need to design questions to which professionals will respond. These questions and answers will aid pupils in completing their enquiry more effectively, thereby learning more. Both public and private industry employees can be of great assistance as long as they have been properly briefed before working with pupils.

For example, in a multi-subject focus topic on road safety in the local area, involving science, design technology and geography, children are asked to identify through fieldwork local pedestrian and traffic blackspots around their school. Then they are asked in groups to design and make models of solutions to one of these blackspots. They brainstorm possibilities, but need to have these confirmed by an 'expert'. A highways engineer explains the different types of solutions and their costings which the department has to consider when making an improvement. The children learn that the safest solution is sometimes the most expensive, by the AOT bringing to their attention the cost factor which they had not considered before. As a result, some of them make compromises to their model solutions, and weigh up the advantages and disadvantages, the costs and benefits, financial, social and in terms of safety.

Citizenship
Key stage 1

- The fire officer, police officer or librarian who visits school or whom pupils may visit on their work site can be used to identify public services which we all depend on.

- Use the study of the local area with the locational knowledge which grows from it – knowing your address, where you live locally and within the UK – to develop the idea of community.
- When studying other places, the similarities and differences concept of geography develops the pluralism concept of citizenship: it is natural for others to speak different languages, have different roles and other differences in their way of life.

Key stage 2
At this key stage, citizenship forms part of geography from the point of view of attitudes, values and issues, and similarities and differences

- Work on any local issue such as a proposition for changes in land use by building a supermarket or more housing, lends itself to developing the concept of democracy. Pupils can collect data on the range of opinions about the proposed development through questionnaires via parents and/or fieldwork interviewing. They can ask the professionals involved for information and see how plans are put to the public and how and when the need for a public enquiry arises. They could also role play a public enquiry simulation.

Health Education
Key stage 1

- Deal with basic road safety relating to local road features or blackspots as one focus in part of a 'Transport' or 'Journeys' topic.
- When pupils visit another environment, discuss the hazards which could affect them in that environment.

Key stage 2

- Develop increased personal awareness of safety issues on fieldwork in different types of environment, for example the top of a castle or by the sea.
- Overlaps between Health Education, Environment Education and Economic and Industrial Understanding are highlighted in developing localities:
  – Encourage pupils to see the cause and effect relationship between health and the lack of service provision in some localities. If there is a lack of doctors, hospitals and sewage disposal in an area, people's health and general expectations are bound to be affected.
  – The availability and management of water and food resources can be compared 'here' and 'there'.
  – Consider the health issues involved in supplying water to pupils' homes and the removal of waste water.

Careers Education

There is considerable overlap with EIU here, because recognising jobs relates to economic activity.

Fieldwork activities described in Chapter 9 develop understanding of ideas about roles and work.

- Map the different work areas of the school and graph the results of a survey to see how many different job types exist in the school.
- Do some fieldwork in the local area to spot different workplaces and associated job types.
- Visit a workplace such as a supermarket and question people about their roles as part of a team. This, of course, would be only part of the focus for the visit.
- Settlement in the local area: from fieldwork experience, maps and a correlated aerial photograph, pupils can work out and map which parts of the area are

predominantly residential and where and what the workplaces are. As part of a local history project, the current situation could be compared to sources of information about the past.

Cross-curricular dimensions

Cross-curricular dimensions are concerned with all aspects of equal opportunities:

- Gender
- Race
- Religion
- Ability
- Culture.

The school's attitude to these should be indicated in the whole-school policy statement. These dimensions can be seen as part of spiritual, moral, social and cultural education.

The OFSTED Framework for the Inspection of Schools has Spiritual, Moral, Social and Cultural Education (SMSC) as one aspect of education to be inspected.

Geography teaching and learning can make very significant contribution to these aspects of a school's life. However, many primary schools do not look to the subject to develop and enhance these aspects. This is unfortunate, as geography has much to offer here. As with cross-curricular themes, half the battle is actually recognising the subject's potential contribution and then highlighting it. The contribution of geography to SMSC is well worth highlighting in your policy.

Spiritual education

Geography is an excellent subject for the occasional highlighting of that transient but essential dimension to any individual's life, a sense of awe and wonder. A child may be at first entranced, but then excited, by his or her sight of a first deep snowfall. Wider landscapes – the view from a hilltop or cliff on a field visit, or a natural feature such as an erupting volcano seen on a video newsclip – may also, skilfully used by the teacher, help pupils to appreciate the amazing nature of the natural world. Pupils who are accustomed to rural or natural environments need reminding of their value. Those who live in inner-city areas with problematic backgrounds need to be made aware that there is an amazing aspect to environments elsewhere which can stimulate them to think beyond the immediate and material one.

It is also possible to wonder at people's better efforts in the built environment. Individual buildings may demonstrate such good design and material qualities that they really enhance their urban environment. Urban design seen from the view of a city settlement on a geography video of a distant place may have striking qualities to which teachers can call children's attention, albeit briefly, to show that humans can work with a sense of pattern on a grandiose scale.

Natural hazards, such as flooding, hurricanes and earthquakes, reflect the enormous and breathtaking power of dynamic physical geography forces.

Moral education

The attitudes, values and issues aspect of geography discussed in Chapter 1 as essential to primary geography has an important part to play in pupils' moral education. The range of attitudes, values and issues generated by a change in land use in the local area, if investigated as part of an enquiry into settlement and environmental change, may help children to see that solutions may be right for some people, wrong for others, but best fit overall, as a democratic compromise needs to be reached. Role-play may be an appropriate strategy to explore issues which are beyond the children's direct experience but of which they have some knowledge or understanding, albeit secondhand.

Global issues, such as rainforest destruction, over-use of other sustainable and non-sustainable resources, environmental and atmospheric pollution, extreme attitudes towards the raising of stock animals for food, are all geographical issues which need exploration in the primary school.

Geography can help to teach the principles which distinguish right from wrong.

Social education

- Does your school take children away on a residential field trip, widening their horizons and requiring them to live as a team?
- Does it invite adults other than teachers (AOTs) from the wider community to work with pupils for short or longer periods of time?
- Do your pupils go out into the community on fieldwork, making observations and collecting data about it, thereby deepening their knowledge of it?
- Does your school, as the result of a geography project, sometimes make its views known to the wider community, for example the highways, planning department or the library?
- Are your pupils expected to work in a variety of ways in their geography work (in pairs, in groups) whilst developing a sense of independence in enquiry work?

All these aspects, which you may develop in the geography curriculum in your school, contribute to pupils' social education. You may not do all of them equally; you may focus on some particularly rather than others; but they should all be encouraging pupils to relate positively to others, to take responsibility, to participate in the community and to develop an understanding of citizenship. It is worth noting that the cross-curricular theme of citizenship is one aspect of the social side of SMSC.

Cultural education

Schools often do not recognise the potential of geography in helping pupils to appreciate the diversity and richness of other cultures. In both key stages, the contrasting locality studies and the requirement to situate localities within their wider context showing how they are linked with elsewhere must be ideal opportunities to present aspects of other cultures to children. The use of key questions such as:

- Where is the distant place?
- How do people live there?
- Why do they do these things/live like that?
- How are their lives changing, and why?
- How do they feel about it? How would we feel about it?

must further children's understanding of different cultures and ways of life.

The use of fiction or factual literature from other countries, in the context of the geography of a locality within it, can enhance learning about the place. Visitors from the countries concerned can stimulate children's questions about those places, as can artefacts from real places.

Multi-cultural education

Although all schools need to consider their policy on cultural development as part of SMSC, multi-cultural education should really be a cross-curricular dimension. All pupils need to be educated to work and live in a multi-cultural environment, and primary geography can be invaluable in promoting tolerance and responsible attitudes towards other races and cultures. As with cultural education, distant-place teaching must highlight not just the culture of particular places, but the range and diversity of races and culture in places all around the world.

For those pupils living in multi-cultural areas, learning with peers whose cultural heritage is part British and part distant places makes learning about those distant places more relevant.

Geography coordinators could well examine their school and class libraries and CD-ROM encyclopaedias for multi-cultural images. Often sufficient quantity and quality of up-to-date resources on people in a range of places are lacking in the geography areas of libraries:

- Material 15 years old or more will provide dated images of peoples and their lifestyles
- Old material is often not presented attractively enough to captivate children's interest
- Black and white photos or artists' impressions are not sufficient
- European, Asian, African and North and South American cultures should be represented. These materials should be regularly monitored and discarded as they become outdated.

It is also worth checking the RE section of libraries. These may be well provided with up-to-date multi-cultural images of people who practise different religions in the UK and abroad and therefore very useful to support learning in geography, too.

For those pupils in schools where multi-cultural contacts are rare, geography is one subject area which brings in the dimension naturally. Studying distant places and learning to value other people's cultures and environments widen pupils' horizons.

The European dimension

As part of the multi-cultural dimension, or separate to it, some schools may wish to highlight this dimension in their school's curriculum. They may possibly teach the beginnings of a modern foreign language or have links with a school in Europe, or are situated in the southern UK where links with or travel to Europe are particularly appropriate. Geography again is an excellent vehicle for promoting the European dimension through the study of themes in European Union localities, regions or countries (see Chapter 9). You could use some basic vocabulary in the modern foreign language to teach orally about the features of your European place on a town plan, for example: *La boulangerie est en face de la gare* – positional geography of basic human settlement features in French, located on a plan.

Just as the different aspects of spiritual, moral, social and cultural education are frequently interlinked, so do cross-curricular themes interlink with cross-curricular dimensions. For example, citizenship is part of social development; environmental education with moral education.

Planning for cross-curricular themes and dimensions

In Chapter 3, the links which geography has with other subjects were explored. Ideally, links with cross-curricular themes should also be explored before planning key stages. The dimensions certainly need to be considered by the staff as part of the whole-school ethos at the level of whole-school policy development.

However, because there is a limit to what we can cope with at any one time, it is suggested that until you are familiar with the themes and know how they relate to the geography parts of a topic, you should plan themes in at the unit of study stage. Ask yourself how the themes relate to the focus questions planned and make a note of which ones you will be dealing with, as shown in the focus-planning examples in Chapter 5.

As you grow in experience, you will find that you are able to see the links more clearly and develop the cross-curricular themes and dimensions more deeply through geography, planning them in explicitly first of all at focus planning level, but later bearing them in mind at key stage planning level when reviewing takes place.

13 RESOURCES

To teach primary geography well a wide range of resources is necessary. Most schools will have some resources, however limited, and in the Appendix a resources audit list for National Curriculum geography, key stages 1 and 2, is provided which will help you to evaluate your current position. Resources which have been referred to in previous chapters will be found listed here.

The resources listed are not the only ones available, but are some that the authors have found useful for classroom teaching. New resources are constantly being produced, and it is recommended that you send for inspection copies where possible, to check for usefulness and value for money. Will the resources be relevant to your key stage plan and its units of work?

Books and resources

Textbooks and schemes

The purchasing of one set of textbooks as a 'scheme of work' to satisfy National Curriculum requirements is not recommended. No one scheme can fit a primary school's individual approach to geography within the context of its own curriculum. However, it is recognised that textbooks can be extremely useful if used as resources in a variety of ways.

1 By using certain pages, sections and chapters of the book as a group or class activity to support planned units of work from the key stage plan.
2 To give teachers information and ideas to use with pupils.
3 To be used by pupils for their own enquiry.

However, there are times when coordinators may find it useful to have a set of textbooks available to guide and support teachers who feel uncertain in this curriculum area, as long as the books are relevant to the appropriate units of work.

Many of the current schemes are being rewritten in light of the Order. Others are being replaced, and many completely new schemes are coming onto the market. Schemes listed here have been published after 1990 unless otherwise indicated.

Schemes

Ginn Key stage 1 and 2 Geography, Bill Chambers and Wendy Morgan, Ginn, 1991.

Keystart (see Atlases below), Collins Longman, 1991.

Oliver & Boyd Key Stage 1 and 2, W.E. and V. Marsden, Oliver & Boyd, 1991.

Collins Primary Geography: World Watch, Key Stage 2, Scoffham, Bridge and Jewson, Collins, 1994.

Oxford Primary Geography Key Stage 2, ed. Steve Harrison, OUP, 1995.

Schoolbase Geography, Stephen Scoffham, Colin Bridge and Terry Jewson, Schofield & Sims, 1986.

Sunshine Geography (part of Sunshine Books key stage 1), Heinemann, 1991

Time and Place, Patricia Harrison and Steve Harrison, Stanley Thornes, 1992.

The Young Geographer Investigates, Terry Jennings, OUP, 1986.

Textbooks for mapwork

Discover Maps, Patricia Harrison and Steve Harrison, Collins Educational/OS, 1988.

Mapping Skills, Tom A. Dodd, OUP, 1985.

Mapskills Activity Book, OUP, 1985.

The Mapskills Atlas, work book and copymasters, Collins Longman, 1992 (top juniors and ideas for staff).

Mapstart Books 1–3, Simon Catling, Collins Longman, revised 1994.

OS Mapstart, Simon Catling, Collins Longman, 1989.

Mapwork 1, David Flint and Mandy Suhr, Wayland, 1992.

Master Maps, Patricia Harrison and Steve Harrison, Collins Educational/OS, 1988.

OS Resource File, Patricia Harrison and Steve Harrison, Collins Educational, 1989.

Philip's Children's Atlas, 1992.

Another related map book:

Moving Into Maps, Heinemann, 1983.

Big books

Various large format (or big books) have been produced specifically for key stage 1 geography, or sometimes for history and geography. Some are an aid to schemes of work, or have accompanying books of photocopiable worksheets or some are atlases in their own right. Useful materials include:

Ginn Key Stage 1 Geography

Folens Geography (age 4-5)

Folens Geography (age 5-7)

Collins Keystart First Atlas (UK and Europe), Collins Longman

Time and Place Key Stage 1 Big Book, Stanley Thornes

The Oxford Infant Atlas Flopover Book, OUP

Teachers' resources for locational work

The following two publications for teachers are extremely helpful, giving clear explanations and plenty of classroom practice ideas for locational knowledge work:
Using Maps and Atlases at Key Stage 2, Stephen Scoffham.
Placing Places (new edition), Simon Catling.
Both of these are published by the Geographical Association (address on page 190).

Atlases

Most of the main publishers have a range of atlases in their lists. Some interesting atlases include:

For young children:

Atlas One, Collins Longman.

Oxford Infant Atlas, OUP, 1993.

First Atlas, Ginn, 1994 (KS1).

The Oxford First Atlas, OUP, 1995.

A First Atlas of the World, Schofield & Sims, 1992.

Rainbow Atlas, OUP, 1995.

All Around the World, Evans, 1995 (KS1 and 2).

Going up the age range for key stage 2:

Philips Junior School Atlas, Philips, 1994.

Atlas 2, Collins Longman 1994.

The World Today, Schofield & Sims, 1992.

The Whole World Now, Schofield & Sims, 1992.

The Oxford Junior Atlas, OUP, 1996.

The Folens OS World Atlas, Folens, 1991.

The Folens OS UK Atlas, Folens, 1994.

A recent trend is to incorporate an atlas and textbook into one. Examples are:

Keystart UK Atlas, Collins Longman, 1991.

Keystart World Atlas, Collins Longman, 1991.

The main atlas publishers are:

Arnold Wheaton
Collins Longman
Heinemann
OUP
Phillips
Schofield & Sims.

There is no perfect atlas. Consider these criteria when choosing:

Audience Is the atlas appropriate for its intended age and ability range?

Function What kind of atlas is it?

Is it a working atlas? Does it have activities, copy masters, outline maps, gazetteer?

Is a range of atlas skills covered?

Publication date When was this edition published?

How up-to-date is it?

Remember that an atlas will probably be out of date as soon as you buy it. Part of the skill in using an atlas is noticing and discussing changed boundaries and altered countries.

Details Do you like a lot of pictures or photographs around the maps?

Do you prefer clear, uncluttered maps?

Do you want a map of the whole of the UK on one page?

Do you want your area of the UK shown fairly large?

Do you want a map which shows southern England or the whole of the UK and northern France and Belgium on the same page to emphasise connections with continental Europe?

Do you want a map which shows the current counties of the UK?

Is the text/size/density of the text appropriate for the intended age range?

Have the maps been oversimplified at the expense of accuracy?

Are the symbols used explained to the pupils?

If abbreviations are used, are they also in the key?

Is the difference between towns, cities and countries clearly indicated by the size difference in the print?

Ease of use Is there a contents page and an index?

Are the page numbers, grid systems or latitude and longitude used to refer to places?

Which of these is most suitable for the age of your pupils?

Projections and orientations Are you told which map projection is being used?

Is it Mercator, Peter's, Eckert, Winkel's (see Glossary)?

Are two or more projections presented on pages for comparison?

Are the world maps all Euro-centred, or are some centred on the North Pole, etc.?

Do aerial photographs and satellite images support the maps?

Images of places How is the sense of place conveyed: by words, through photographs, through sketch drawings?

Do the pictures reflect contrasts within the country or continent?

Are the stereotypical images challenged or reinforced? (For example desert = camel; tundra = igloo.)

Are people of both genders from varying ethnic and cultural backgrounds shown in a range of roles?

Map clarity Does each map need and have: a title, a key, a scale, a compass pointer?

Are both metric and imperial units of measurement shown, or only one?

Do foreign place names have English or native language spelling, for example Brussels/Bruxelles? Which do you require?

Is the colour key helpful? For example does green show lowland or vegetation?

Cost Is the cost reasonable?

Format and durability Is the atlas attractive to children?

How durable is it?

Look at the binding, cover lamination, etc.

Geographical Work in Primary and Middle Schools, Appendix 4, has a more detailed set of criteria to which you may wish to refer when choosing an atlas.

Maps and plans

When using materials to support mapwork a major resource will be that of maps and plans. The following list shows the variety.

- Street maps
- Road maps
- Postcard maps
- Maps on stamps
- Maps in adverts
- Road sign maps
- Housing estate maps
- Town centre maps
- Tourist maps
- Trail maps
- Ordnance Survey maps
- Bus maps
- Rail maps
- Underground maps
- Building plans
- Room plans
- Board game maps
- Guide book maps
- Picture maps
- Atlas maps
- 'Antique' maps
- Microchip circuit 'maps'
- Wall chart maps
- Walkers' maps
- Land use maps
- Sketch maps
- Globe maps
- Architects' plans
- Developers' plans
- Maps in stories

The types of OS maps needed and the key stage to which they are appropriate has been described in detail in Chapter 8.

Mapwork General details about sources for OS maps were also given in Chapter 8. Information and price lists on all OS maps and wall charts can be obtained from:

Education Team
Ordnance Survey
Romsey Road
SOUTHAMPTON
SO16 4GU
Tel: (01703) 792960
Fax: (01703) 792039

The National Map Centre
22–24 Caxton Street
LONDON
SW1H 0QU
Tel: (0171) 222 2466
and Stanfords, listed under Globes.

Ordnance Survey now have many listed regional and local suppliers for larger scale maps as well as smaller scale ones. Check their catalogue for your nearest supplier.

A full range of OS products is also available through Chas. E. Goad, as well as their own town centre plans. Goad plans are accurate plans of shopping centres and high streets. These plans show in detail the layout of the shops, the retailer and what they trade in. They are updated, some annually and others once every two years, so comparisons can be made of the shops now, when the plan was drawn and some number of years ago. It is possible to buy the copyright for £2, which enables the buyer to reproduce parts of the plans for pupils' worksheets. There is also a primary pack to help teachers develop work in shopping centres using the plans (for ages 7 to 11) – *Education Pack A: Goad Shopping Centre Plans in Junior Schools.*

Goad also produces an excellent catalogue entitled *Charles Goad Mapping Division: Education Information Pack* which you can obtain free of charge. It gives clear information about a wide range of maps, atlases, globes, wall maps and CD-ROMs.

Chas. E. Goad
8–12 Salisbury Square
OLD HATFIELD
Hertfordshire
AJ9 5BJ
Tel: (01707) 271171
Fax: (01707) 274641

Globes

Globes are either physical – showing land forms – or political – showing countries. They can be bought from many stationers, for example W.H. Smith, high street shops such as the Early Learning Centre, and through educational catalogues, for example Phillips, Hestair Hope, E.J. Arnold.

Stanfords has an extensive globe catalogue:

Stanfords,
12–14 Long Acre
LONDON
WC2E 9LP.
Tel: (0171) 836 1321

If you are buying a hard globe, it sometimes pays to buy one with a metal arm, as these last longer with children leaning on them. Good-quality globes are expensive; inflatable globes provide a viable and inexpensive alternative. They start around £4, and can be bought from Chas. E. Goad and:

Cambridge Publishing Services
PO Box 62
CAMBRIDGE
CB3 9NA

Other equipment

Compasses, clinometers and weather-measuring instruments can be obtained through educational suppliers. For more specific help on weather studies, including weather forecast logging maps, contact:

The Meteorological Office
Marketing Services
Sutton Building
London Road
BRACKNELL
Berks
RG12 2SZ
Tel: (01344) 854 818

Collect your own rocks and fossils. Commercial packs are available, for example Primary Core Pack (key stage 2) from:

Earth Works
Geography Supplies Limited
16 Station Road
Chapeltown
SHEFFIELD
S30 4XH
Tel: (0114) 455746

Oblique aerial and side view photographs sources

Several companies now publish high quality, durable photograph sets which are A4 format or larger. Such photos relate to a range of places and themes, and show physical and human geography features and activities. They make good resources about which pupils can raise geographical questions relating to patterns and processes. Teachers' notes usually accompany these packs. Particularly good packs are as follows:

Topic Packs
Weather
Water
Climates
Landscapes
Polar Regions
Deserts

Cost: £12 per pack. Available from:

Philip Green Educational Ltd
112A Alcester Road
STUDLEY
Warwickshire B80 7NR
Tel: (01527) 854711
Fax: (01527) 854385

Folens Photo Series
Transport
Aerial Photos
Physical Features
Weather

Cost: £11.99 per pack. Available from:

Folens Publishers
Albert House, Apex Business Centre
Boscombe Road
DUNSTABLE
LU5 4RL
Tel: (01582) 472788
Fax: (01582) 472575

Collins Primary Geography World Watch Themes Pack (Key Stage 1)
Scoffham, Bridge and Jewson, Collins Educational, 1993

Vertical aerial photograph sources

Sources Local newspapers have aerial photographs, usually in black and white.

Local flying clubs may have members who are prepared to overfly your area and take photographs if you cover costs.

Local flying centres may be approached for a hire service and local hot air balloon enthusiasts/clubs can be hired or paid costs.

Commercial aerial photograph companies offer varying packages. Check that their service is a colour one, and that they will laminate photographs. This service costs extra, but extends the use and life of photographs and is well worth it.

Reputable companies The National Remote Sensing Centre, Air Photo Group has an education officer. INSET sessions can be provided for primary teachers. Air Photo's speciality is vertical aerial photos.

NRSC Air Photo Group
Arthur Street
BARWELL
Leicestershire
LE9 8GZ
Tel: (01455) 844513

Photoair has packs on rivers: the Thames, Severn, and your regional river.

Photoair
191A Main Street
Yaxley
PETERBOROUGH
PE7 3LD
Tel: (01733) 241850

Hunting Aerofilms
Gate Studios
Station Road
BOREHAM WOOD
Herts
WD6 1EJ
Tel: (0181) 207 0666

Ask your local secondary school for its discarded GCSE oblique and vertical aerial photographs. You may be given sets, although they are unlikely to be of your area.

Specific books for the school library which are suitable for infants and juniors include:

The Aerofilms Book of Britain from the Air
The Aerofilms Book of England from the Air
The Aerofilms Book of Scotland from the Air
The Aerofilms Book of Ireland from the Air
Yorkshire from the Air
The Changing Face of Britain
(available from Hunting Aerofilms);

Above London, Editors: Robert Cameron and Alistair Cooke, Andre Deutsch, and
Above Paris, Salinger, Andre Deutsch are both obtainable from bookshops.

Satellite images

These are useful resources to motivate, challenge and stimulate children relating to themes or place work. Regional (e.g. SE England), UK, Europe and world satellite images are now widely available in display-size format. They are best bought already laminated: the extra cost is readily regained with longevity and practicality of use. Lamination allows for wipe-off pen and sticker use.

Such satellite posters can be obtained from Stanfords, listed earlier under 'Globes' or by mail order from:

MJP Geopacks
PO Box 23
St Just
PENZANCE
Cornwall TR19 7JS
Tel: (01736) 787808

A specific resource for teachers and pupils at key stage 2 now exists:

Images of Earth: A Teacher's Guide to Remote Sensing in Geography at Key Stage 2, Barnett, Kent and Milton (eds), Geographical Association, 1994

It contains colour images for classroom use. Some of the images relate to St Lucia in the Caribbean and Castries, a commonly used locality there, found in a whole range of other media now.

Multiple copy packs of images of Manaus, Brazil (Amazon River), Mt St Helen's, USA and Castries, St Lucia from the pack are also available separately.

Information technology

The range of software available for National

Curriculum geography has grown considerably over the past few years. CD-ROM resources are an area of particular growth.

All open-ended software available for communicating and handling information continues to be of prime importance for supporting geography. Concept keyboards continue to be useful for promoting geographical vocabulary and the notion of routes.

A wide range of geography-specific software has been developed recently. It needs to be evaluated for its genuine usefulness in developing geography concepts and skills. Much of it is skills reinforcement-based or is better done through real practical teaching. The best geography-specific software involves OS map extracts, sometimes combined with photos and other data or information, or develops concepts of landscapes through three-dimensional block diagrams or models. Such software can assist pupils with special educational needs who have learning difficulties. In some cases, they could also extend brighter pupils. CD-ROMs, such as the *Distant Places Interactive Atlas* (AV Enterprises Ltd) and *Windows on the World: Primary CD Atlas* (Nelson and YT/ILP), are of particular use in primary schools. See Figure 13.1 for further information on software available.

Several recent publications from NCET are helpful in promoting the integration of IT with geographical work. They are: *Focus on IT and Primary Humanities*, *IT in the National Curriculum*, and *Approaches to IT Capability, Key Stages 1 and 2*. In line with the Revised Order is the joint publication from NCET and the GA, *Primary Geography: A Pupil's Entitlement to IT* (see Chapter 3, page 33). The whole guidance paper is worth obtaining as it gives much straightforward practical help. NCET's address is:

NCET
Millburn Hill Road
University of Warwick Science Park
COVENTRY
CV4 7JJ
Tel: (01203) 416994

The guidance paper can also be obtained from the Geographical Association (address page 190).

Another useful address for IT is:
Advisory Unit for Microtechnology in Education
Endymion Road
HATFIELD
Hertfordshire
Tel: (01707) 265443

Media and audio visual aids

Media Educational broadcasts are a very useful resource for geography work. Both television and radio programmes for schools come with professional and usually detailed teacher's notes and often with relevant worksheets.

Programmes should be used as a resource, not the rationale, for the unit of work. Television can help particularly in the provision of up-to-date material on distant places, both UK and wider world. Watch out for programmes which help locality studies. Ideally teachers should preview programmes.

- Programmes should be recorded to provide a resource bank and to facilitate appropriate use.
- It may only be appropriate to draw on one or two programmes from a series. Take care to choose only those which supplement or form a focus for some part of pupils' work. Sequences can be chosen from a programme to make a particular point.

Figure 13.1

Geographical software

Software type	Package	Company	Availability
Word processing packages	Pen Down	Longman Logotron	A
	Desktop Folio	ESM	A
	Folio	ESM	BBC
	Stylus	Mape	BBC, N
	Magpie	Longman Logotron	A
	Caxton Press	Newman College	480Z
	Front Page Extra	Newman College	A, BBC, N
	Caption		BBC
	Developing Tray	Inner London Education Computing Centre	
Communication with other schools	Email		
	Fax Machines		
Data handling packages	Our Facts, Sorting Game, Notice Board, Datashow, Branch	NCET data handling pack	BBC, N, 480Z
	Notice Board	Newman College	BBC
	Grass		BBC, N
	Grasshopper		BBC, N
	Datasuite		A
	Graph It		BBC
	Touch Explorer Plus	NCET	BBC, N
Directional packages	Logo (version 1)	Longman Logotron	A, BBC
	Logo (version 2)	Research Machines	N
	Tiny Logo	Topologica	A, N
	Turtle	Valiant or Jessop	A, BBC, N
	Roamer	Valiant	freestanding
	Pip	through Fernleaf from Swallow Systems	freestanding
Data logging packages	First Sense	Philip Harris	
	Weather Reporter	Advisory Unit for Microtechnology in Education	
	Sensing Science Pack	NCET	
Software designed to support geography specifically (key stage 2)	Mapventure	Sherston Software	BBC, N
	Viewpoints	Sherston Software	A
	Mapping Skills	ESM	BBC
	SMILE maths	Inner London Education Computing Centre	
	Whatley Quarry	NCET	
	List Explorer	NCET	BBC Master
	World Atlas	Software Toolworks	
	World Wise	Bourne Educational	
			BBC
			(A = Archimedes) (BBC = BBC) (N = Nimbus) (480Z = Research Machines)

- Often the best use of a programme is made by seeing or listening to the programme once then replaying key sequences to reinforce an idea. Pupils can interact with the material, having been guided by the teacher to focus on particular aspects or answer specific questions.
- Small group viewing and listening can be used for further differentiation.
- Don't forget that the pause or freeze-frame button can be very useful in clarifying parts of the programme.

Useful programmes

- *Our World*, key stage 1, ITV
- *Geography – Start Here*, key stage 1 and 2, ITV renamed *Geography 7–11* in 1995/6
- *Search*, key stage 2, ITV
- *Environments*, key stage 2, ITV
- *Going Places*, key stage 2, ITV
- *Watch*, key stage 1, BBC
- *Landmarks*, key stage 2, BBC
- *Zig Zag*, key stage 2, BBC
- *Geography 7–11*, key stage 2, Channel 4

Individual programmes from a wider range of series can complement your work; check the programme titles and synopses.

Audio visual aids BBC's *Teaching Today* series, prepared for in-service training, has issued a video comprising two programmes on key stages 1 and 2 primary geography. There is a further one on Environmental Education in primary schools. The videos have an accompanying booklet to help head teachers or coordinators who wish to use them with the whole school staff for school focused staff development.

You can borrow or hire videos from many of the organisations already listed. The *Environmental Review* video from Ark is very useful for upper juniors. Filmstrips and slide sets can be purchased from similar organisations. Remember to check the age range for which they are designed before ordering.

Ark
489–500 Harrow Road
LONDON W9
Tel: (0181) 968 6780

Videos entitled 'Rivers' and 'Settlement' are now available from:

Gogglebox
Resources for Learning Ltd
19 Park Drive
Heaton
BRADFORD
West Yorkshire BD9 4DS
Tel: (01274) 544155
Fax: (01274) 549391

at £7.50 including postage and packing.

Useful addresses for UK studies

Conservation Trust
National Centre for Environmental Education
George Palmer Site
Northumberland Avenue
READING
RG2 7PW
Tel: (01734) 868442

Council for Environmental Education
School of Education
University of Reading
London Road
READING
RG1 5AQ
Tel: (01734) 318921

Council for the Protection of Rural England
Warwick House
25–27 Buckingham Palace Road
LONDON
SW1W 0PP
Tel: (0171) 9766433

Countryside Commission
John Dower House
Crescent Place
CHELTENHAM
GL50 3RA
Tel: (01242) 521381

English Heritage
Fortress House
23 Saville Row
LONDON
W1X 2HE
Tel: (0171) 9733000

Field Studies Council
Preston Montford
Montford Bridge
SHREWSBURY
SY4 1HW
Tel: (01743) 850674

Forestry Commission
231 Costorphine Road
EDINBURGH
EH12 7AT
Tel: (0131) 3340303

Learning through Landscapes
Third Floor
Southside Offices
The Law Courts
WINCHESTER
Hampshire
SO23 7DU
Tel: (01962) 846258

National Association for Environmental Education
Wolverhampton Polytechnic
Gorway Road
WALSALL
WS1 3BD
Tel: (01922) 31200

National Trust
36 Queen Anne's Gate
LONDON
SW1H 9AS
Tel: (0171) 2229251

National Conservancy Council
Northminster House
PETERBOROUGH
PE1 1YA
Tel: (01733) 40345

Ramblers Association
1–5 Wandsworth Road
LONDON
SW8 2IJ
Tel: (0171) 5826878

Royal Society for the Protection of Birds
The Lodge
SANDY
Bedfordshire
SG19 2DL
Tel: (01767) 80551

Soil Association
86–88 Colston Street
BRISTOL
BS1 5BB
Tel: (0117) 290661

Tidy Britain Group
The Pier
WIGAN
WN3 4EX
Tel: (01942) 824620

Town and Country Planning Association
17 Carlton House Terrace
LONDON
SW1Y 5AS
Tel: (0171) 9308903

Woodland Trust
Autumn Park
Dysart Road
GRANTHAM
Lincolnshire
NG31 6LL
Tel: (01476) 74297

Useful addresses for distant places studies

Action Aid
Hamlyn House
Archway
LONDON
N19 5PG
Tel: (0171) 2814101

For publications:
Action Aid
Chataway House
CHARD
Somerset
TA20 1FA
Tel: (01460) 62972
Fax: (01460) 67191

British Red Cross
9 Grosvenor Crescent
LONDON
SW1X 7EJ
Tel: (0171) 2355454

Catholic Institute for International Relations
22 Coleman Fields
LONDON
N1 7AF
Tel: (0171) 3540883

Catholic Fund for Overseas Development
2 Romero Close
Stockwell Road
LONDON
SW9 9TY
Tel: (0171) 7337900

Centre for Alternative Technology
Llwyngwern
MACHYNLLETH
Powys
SY20 9AZ
Tel: (01654) 702400 and 703409
(Education Office) 703743
(Bookshop) 702948

Centre for Global Education
Longwith College
University of York
YORK
YO1 5DD
Tel: (01904) 413267

Centre for World Development Education
Regent's College
Inner Circle
Regent's Park
LONDON
NW1 4NS
Tel: (0171) 4877410

Centre for World Development Education
1 Catton Street
LONDON
WC1R 4AB
Tel: (0171) 4877410

Christian Aid
PO Box 100
LONDON
SE1 7RT
Tel: (0171) 6204444

Commonwealth Institute
Kensington High Street
LONDON
W8 6NQ
Tel: (0181) 6034535

Council for Education in World Citizenship
Seymour Mews House
Seymour Mews
LONDON
W1H 9PE
Tel: (0171) 9351752

Development Education Centre
Gillott Centre
998 Bristol Road
BIRMINGHAM
B29 6LE
Tel: (0121) 4723255

Latin American Bureau
1 Amwell Street
LONDON
EC1R 1UL
Tel: (0171) 2782829

National Association of Development Centres
8 Endsleigh Street
LONDON
WC1H 0DX
Tel: (0171) 3882670

National Geographic Society
PO Box 19
GUILDFORD
Surrey
GU3 2NY
Tel: (01483) 33161

Oxfam
274 Banbury Road
OXFORD
OX2 7GZ
Tel: (01865) 56777

Save the Children
Mary Datchelor House
17 Grove Lane
Camberwell
LONDON
SE5 8RD
Tel: (0171) 7035400

Scottish Catholic International Aid Fund
5 Oswald Street
GLASGOW
C1 4QR
Tel: (0141) 2214446

Scottish Education and Action for Development
29 Nicolson Square
EDINBURGH
EH8 9BX
Tel: (0131) 6670120

UNICEF UK
55–56 Lincoln's Inn Fields
LONDON
WC2A 3NB
Tel: (0171) 4055592

Welsh Centre for International Affairs
Temple of Peace
Cathays Park
CARDIFF
CF1 3AP
Tel: (01222) 384912

Worldaware
1 Catton Street
LONDON
WC1R 4AB
Tel: (0171) 8313844
Fax: (0171) 8311746

World Education Development Group
29 Watling Street
CANTERBURY
Kent
CT1 2UD
Tel: (01227) 766552

World Wide Fund for Nature (WWF)
Panda House
Weyside Park
GODALMING
Surrey
GU7 1XR
Tel: (01483) 426444
Fax: (01483) 426409

Distant place packs

Many of these packs, which include quality A4 photos, teachers' notes and pupil information, are now linked with videos. Some of them can be bought with videos. Others can be purchased in addition to or separately from the video, which is part of an educational television geography programme. The range of these resources is increasingly wide and good, especially for localities beyond the UK. Most packs are produced with key stage 2 in mind, but imaginative teachers successfully adapt them for key stage 1. Most need some modification, even at key stage 2, to suit your school and pupils. Some schemes of work now deliberately include localities which are similar to those found in published packs. For example, Chembakolli, St Lucia, Change in the Swat Valley have pages in the *Ginn Key Stage 2 Geography* scheme.

Localities in economically developing countries: Specific location packs

South America

1 *Pampagrande: A Peruvian village (1992)*
This pack focuses on a rural locality and could be used to fit in with topics on Family, Farming, Shopping, etc. Useful teachers booklets with background information. Produced by Action Aid.

Cost: slide set £15.00. Available from Action Aid, Chard (address on page 180).

Africa

2 *Where camels are better than cars: Dourentza, Mali, West Africa (1992)*
Focuses on the lives of four people who each come from different cultural groups. Includes 34 colour photographs, a poster, plan of Dourentza market, a teacher's book and a pupil's book.

Cost: £13.00, plus £2.28 for postage and packing. Available from Development Education Centre, Birmingham (address on page 180).

A video has been produced to accompany the pack, increasing the price to £25.00. The video uses video stills, a technique to which teachers and children can adjust successfully. The materials are very highly regarded by teachers who have used them.

3 *Cairo: Four Children and their City: A Video Pack for Key Stage 2 (1995)*
Cairo is the largest city on the African continent, and has been a cultural crossroads for thousands of years. This video pack provides a 20-minute colour video, 'Welcome to Cairo!', a set of colour photocards, a resources book of information and classroom strategies for teachers, and photocopiable teaching resources all integrated to the requirements of the National Curriculum

Through the eyes, words and activities of four children with different backgrounds and from different parts of Cairo, the video introduces us to Islamic culture, the conflicts over space in a city, the role gender plays in defining lifestyles, the impact of traffic and the modes of travel in a city, and to the contrasts in lifestyle within the unified area of a city.

The photocards extend the themes beyond the video and add a further dimension to the study of Cairo.

The resources book has been developed by a group of teachers in Hampshire working with Dylan Theodore in the Oxfam Education unit in Southampton. It includes photocopiable materials for pupils and classroom-tested strategies for delivering the National Curriculum through the resources in the pack.

Cost: £25.50, plus £5.10 for postage and packaging. Obtainable from:

Oxfam Publishing
c/o BEBC Distribution
PO Box 1496
POOLE
Dorset
BH12 3LL

4 *Kano: Discovering Aerial Photographs 11, Upper Juniors*
An innovative pack including 7 colour aerial photographs, an audio tape and 12

photo sheets plus supporting pupils' and teachers' materials on Kano in Nigeria.

Cost: £37.50, plus £3.95 for postage and packing. Available from:

National Remote Sensing Centre
Airphoto Group
Arthur Street
BARWELL
Leicester
LE9 8GZ

5 *Living and Learning in a Tanzanian Village: A child's perspective, Kirua, Tanzania (1992)*
Photo pack (black and white) and activities, using the case study of Kirua Primary School in Tanzania. Emphasis on enabling pupils to make links with their own school lives and locality. Goes a long way in showing how links can be made with a local study. In spite of black and white photos, this is a good pack. Produced by Manchester Development Education Project.

Cost: £11.00. Available from Oxfam Education in Brighton.

6 *Kapsokwony: Rural Kenya (1992)*
This pack provides a case study of a rural community in Kenya. 30 colour photos with teachers' information books on Kenya and Kapsokwony and a range of other information sheets and resources. Produced by Action Aid.

Cost: £15.00. Available from Action Aid, Chard (address on page 180).

7 *Nairobi: Kenyan City Life (1992)*
One of the few packs focusing on an urban locality. The locality studied is a slum area on the outskirts of the city but much of the pack looks at central Nairobi and there are plenty of photos showing a modern, bustling city as well. Produced by Action Aid.

Cost: slide set £15.00. Available from Action Aid, Chard (address on page 180).

8 *Palm Grove: Zambia (1992)*
Mixture of colour and black and white photos. Large poster with map of Africa and photos of 'Education in Zambia'. Teacher's booklet useful for putting Zambia in economic and political context. Focuses on families, school, weather, water, food, journeys, tourism, trade and work. Produced by UNICEF UK.

Cost: £11.95, plus £1.79 for postage and packing. Available from UNICEF UK (address on page 181).

9 *Addokorpe: Life in a Ghanian Village (1992)*
28 good colour photos plus book of teacher's notes with exercises (rather than activities) and maps. Information on homes and buildings, food, work, children, weather and links with other places. Produced by World Vision and Worldaware.

Available from Worldaware (address on page 181).

10 *Kaptalamwa: A Village in Kenya (1994) Maureen Weldon*
Kaptalamwa, close to the equator and with an estimated population of 1500, forms the basis of this study. The photopack contains 28 full-colour A4 photographs, over 60 pages of notes, including 18 A4 activity sheets, and numerous maps and black and white photographs of the area

Information about the village, its customs and way of life, statistics on weather and climate, and background information on Kenya are also included. The activity sheets, specific to key stages 1 and 2, utilise resources both from the pack itself and elsewhere and form the basis of the studies. Produced by the Geographical Association.

Cost: £12.00 (members) and £18.00 (non-members). Available from the Geographical Association (address on page 190).

11 *Baricho: A Village in Kenya (1992)*
Case study of a village in central Kenya, specially selected and prepared to serve as an example of a locality in an economically developing country for key stage 2. Information and activities on the village and the life of an individual family, together with background on Kenya. Many charts and maps included, together with pupil activity sheets, all photocopiable for class use, plus a set of colour photographs. Produced by Warwickshire World Studies Centre.

Cost: £13.50, plus £1.55 for postage and packing. Available from:

Development Education Dispatch Unit
153 Cardigan Road
LEEDS
LS6 1LJ
Tel: (0113) 2784030

12 *Kenya*
Resource pack which includes maps, photographs, simple statistics and promotes artefact use. It can be used alongside the BBC Schools TV Zig Zag Unit on Kenya, or independently support the study of an economically developing country and locality. It contains a poster, 17 photocards, 12 activity sheets and teacher's notes. Johnathan Barnes, a Kent primary headteacher, assisted and advised with the pack.

ISBN 0563 397 365. Available from:

BBC Educational Publishing
BBC White City
201 Wood Lane
LONDON
W12 7TS

13 *The Gambia: Isatou, Chloe and You: A Comparative Study of Three Children and Three Localities, Key Stages 1 and 2 (1994)*
This multi-media pack enables pupils to compare a family in a locality in the Gambia with a family in Marlborough in the UK. In addition, pupils are asked to use their own life and locality to make comparisons. The pack has developed from a longstanding links between Marlborough and the Gambia. It includes 24 A4 colour photos, teachers' information and notes, maps, worksheets and a sound tape relating to both localities. An optional but very worthwhile video can also be purchased.

Cost: pack price without video £21, plus £3 for postage and packing; with video £29, plus £4 for postage and packing. Available from:

Wiltshire World Studies Centre
St John's School
Choppingknife Lane
MARLBOROUGH
Wiltshire
SN8 2AO
Tel: (01672) 514078

India and Indian sub-continent

14 *Chembakolli: A Village in India (1991) Key Stage 1 and 2*
Includes teacher's notes and booklet, information booklet, 30 A4 colour photos, pupil activities and a map. Can be successfully adapted for infants. Produced by Action Aid.

Bangalore: Indian City Life (see **18** below) makes a good contrast to this pack ensuring that pupils don't think all Indians live in villages.

Cost: slide set £15.00. Available from Action Aid, Chard (address on page 180).

15 *Ladakh: North India (1992)*
This pack focuses on the Tibetan Children's Village School in Northern India. There are 28 colour photos and an extensive set of sheets with background information and activities. Because this pack is about an exiled group, it could form part of a topic on

journeys or migration as well as a locality study. Produced by the Geographical Association.

Cost: 12.00 (members), £18.00 (non-members). Available from the Geographical Association, Sheffield (address on page 190).

16 *Neighbours: The Life and Times of Yesudas Keknel: Anand Gram, North India (1992)*
A video pack providing a case study of a young boy. Includes information cards, booklet and maps. The pack aims to show the way of life in a suburban settlement in India, to show the similarities and differences between Anand Gram and the pupil's local area, and to show how a community has worked together to overcome poverty. About the life of a young boy and family in Delhi, it compares British and Indian cultures. It features community life, schools, public events and local industry. Produced by Action Aid, this video won first prize in the 1993 Education Television Awards.

Cost: 30-minute video pack £25.00. Available from Action Aid, Chard (address on page 180).

17 *Savituri: Pillayanatha, Tamil Nadu, South India*
A 15-minute video with teacher's booklet telling the story of 12-year-old Savithri and her friends who are making changes to their lives. Although focusing on a poor rural community, the children in the video show how they are taking charge of their own lives. Produced by Christian Aid.

Cost: £9.99. Available from Christian Aid (address on page 180).

18 *Bangalore: Indian City Life – A Locality Case Study*
Lifestyles ranging from a middle-class family to homeless street children are examined through photos, sketch cards and an adventure booklet. Produced by Action Aid. A good contrast to Ladakh (**15**) and Chembakolli (**14**) packs.

Cost: colour photopack £15.00, slide set £15.00. Available from Action Aid, Chard (address on page 180).

Pakistan

19 *Change in the Swat Valley, Pakistan. A Locality Case Study*
Pupils investigate the social and economic changes taking place in a traditional rural community. Includes an activity on the physical changes of the Swat river valley. Complements the BBC Landmarks series on Pakistan. Produced by Action Aid.

Cost: colour photopack £15.00. Available from Action Aid, Chard (address on page 180).

20 *Change in Pakistan*
A set of four A3 colourful posters showing change in rural Pakistan. Produced by Action Aid.

Cost: £3.00. Available from Action Aid, Chard (address on page 180).

21 *A Tale of Two Cities: London and Calcutta*
A photopack, good for home region/distant place, or contrasting locality in the UK/contrasting locality in an economically developing country, similarity and difference work. It is also good for rural/urban contrast work, for example to contrast with the *Village in India* pack. It is aimed at the 5 to 9 age range.

Available from the World Wide Fund for Nature (address page 181).

The Caribbean

22 *Focus on Castries: St Lucia, West Indies (1992) Key Stage 1 and 2*
Looks at teaching about localities via a family focus. 28 colour photos divided into

three sections: Growing up in Castries, Environment, and Bananas. Photos show variety in wealth in small, urban locality. Produced by the Geographical Association and Worldaware.

Cost: £11.00 (members). Available from the Geographical Association (address on page 190).

A whole host of complementary resources (listed below) exist to make St Lucia a well-resourced place.

a) *Key Stage 1 and 2. BBC Radio – St Lucia Notes*
A comprehensive, 24-page set of notes about Castries, St Lucia, which accompany the BBC Radio Series. Background notes, activities and some resources (maps, photos, recipes, statistics, etc.) are included for each programme.

Cost: £3.00.

b) *Key Stage 1 and 2.* Primary Geographer, *No. 9, St Lucia issue (1992)*
A special issue of the primary school journal produced by the Geographical Association. Its focus is on teaching about a locality in an economically developing country, using Castries in St Lucia as an example.

Cost: £4.00 (members), £6 (non-members). Available from the Geographical Association (address on page 190).

c) Primary Geographer, *No. 22, St Lucia special (1995)*
An excellent update. Cost: £5.00 (members), £7.50 (non-members). Available as (b).

d) *At Home in St Lucia (1992)*
The Harvey family live in Gros Islet, north of Castries. This family, also the focus of other classroom material about St Lucia, are featured at home, work, school and play. 12 colour slides with full notes.

Cost: £9.50 (order no. S-101). Available from Worldaware (address on page 181).

e) *Lessons on Castries*
A booklet of 20 lessons using available resources on St Lucia for key stage 2. The lessons develop theme, place context, locality and skills work.

Cost: £4.00 (members), £6.00 (non-members). Available from the Geographical Association (address on page 190).

f) *Banana Landscapes (1992)*
A slide set focusing on agriculture in St Lucia. The cultivation of bananas is examined, from the banana tree in the field, through the packing and washing processes and to the port. 12 colour slides with full notes.

Cost: £9.50 (order no. S-100). Available from Worldaware (address on page 181).

g) *Video Pen Pals: St Lucia*
A 15-minute video broadcast as part of Channel 4 programme about growing up in St Lucia, focusing on the lives of children in Soufriere.

Cost: £6.59. Available from Worldaware (address on page 181).

22 *Jamaica Resource Pack*
Contained within a plastic folder, 20 A4 laminated photographs and a 32-page pupils' book are included in this pack. The pupils' book includes background information, maps and data, as well as evocative photographs depicting life in Jamaica. Pupils are encouraged to view the case studies and compare and contrast with their own local areas. This pack supplements the Going Places series on Jamaica.

Cost: £9.95 (IPC 123957). Available from:

The Educational Television Company
PO Box 100
Warwick
CV34 6TZ

Contrasting UK locality resources

If you are unable to link with another school and exchange materials, both commercial and pupil-produced, or to visit a contrasting locality, preferably on a residential visit for upper juniors, then these photo and data packs are available. Most of the packs refer to rural localities as most British schools are urban. Thus there is a lack of urban locality resources.

1 *Flatford: A Contrasting UK Locality for Key Stages 1 and 2 (1994), Edward Jackson and Wendy Morgan*
Flatford Mill, world-famous for its Constable connections, is also a centre for residential fieldwork and ideal for the study of a contrasting UK locality. This multi-faceted new resource comprises:

- OS map extracts of five different scales, three in full-colour and two with multiple copies, plus sketch maps
- 30 full-colour A4 photographs, including several aerial shots
- Data banks on people, goods, services, weather
- Full background information
- 80 graded activities.

Cost: £20 (members) £30 (non-members). Available from the Geographical Association (address on page 190).

2 *Discover Godstone*
An exciting new resource pack providing teachers with everything they need to carry out an in-depth study of a contrasting UK locality. Based on the children and village of Godstone in Surrey, it has been designed for KS1 and 2 and written by primary school teachers.

Cost: £19.95, plus £2.25 for postage and packing. Available from the National Remote Sensing Centre (address on page 174).

3 *Malham Tarn Resource Pack and Video*
This pack includes 27 A4 sheets of information for pupils, maps and diagrams, plus aerial photographs, pictures and Yorkshire Dales National Park leaflets to give a complete picture of the area.

Cost: resource pack £9.95 (IPC 123861). Available from the Educational Television Company (address on page 186).

The video is divided into five sections:

- Around Malham Tarn
- Hills, valleys and streams
- Farms and farming
- Malham village and the National Park
- Activities and facilities.

Teachers will find the video a useful aid in the classroom whether or not they are planning a visit to Malham, especially if it is used in conjunction with the resource pack. Together they provide the materials needed to study Malham as a contrasting locality.

The first four sections show useful examples for physical and environmental geography. The last section shows what it will be like for the children that actually visit the area, and covers issues that would interest parents such as accommodation, health and safety.

Cost: 25-minute video £14.95. Available from:

Cambridge Video Production
University of Cambridge Local
Examinations Syndicate
1 Hills Road
CAMBRIDGE
CB1 2EU
Tel: (01233) 553311 (extension 3416)

The video and pack are also available together at a price of £22.50.

4 *Betws-y-Coed: A Contrasting Locality in Wales (1995)*

A well-produced pack with a clear structure for enquiry, containing A4 photos, postcards, teacher's notes and clear appropriate worksheets. Better supplemented with your own resources if you have been there. Prepared for KS2 but with possibilities for KS1.

Cost: £14.75, plus £1.00 for postage and packaging. Available from:

Field Studies Council Publications
Preston Montford
SHREWSBURY
SY4 1HW

5 *A Rural Locality: Sedbergh (1995)*
This small town in the Yorkshire Dales is made accessible for pupils through a range of maps, posters, photos accompanied by an extensive teacher resource guide.

Cost: £49.99 (Folens ref. F6298). Available from Folens (address on page 174).

6 *Geography Study Kit: Plymouth – A Waterfront City*
ISBN 0590 531379

7 *Geography Study Kit: Eyam*
ISBN 0590 531786
These key stage 2 packs, both published by Scholastic, are extremely detailed and comprehensive, but are most effectively used if you are either taking pupils to the areas on a residential field trip or you know the areas well yourself and have your own photos, slides, artefacts, etc. The visual resources provided need supplementing to permit the most motivating use of the packs.

Cost: £27.50 each. Available from Scholastic.

8 *The Local Network Resource Pack (Complements the Channel 4 TV programme)*
This pack is designed to help children study a contrasting UK locality. It contains maps, notes, photographs, children's work and other material from three different places featured in the programmes: Colton, a village in Staffordshire; the Old Dean Estate in Camberley, Surrey; and Swansea city centre.

Although this is a useful pack especially when used with the video programmes, consider the issues raised by looking at four different localities. Is it better to focus on one and get some depth? This pack could now be usefully used to develop theme work in the UK context, using the localities as illustrative material, rather than studying four localities in insufficient depth.

Cost: £12.95 (IPC 123959). Available from the Educational Television Company (address on page 186).

European resources

These resource packs written about European Union localities can still be used in this context. Most of them contain details about the region and country in which the place is situated. All can be used for relevant theme work.

1 *European Locality Pack: France: Wasquehal, near Lille, Key stage 2*
This pack, published by Ginn, is centred on Wasquehal, a suburb of Lille. It contains teacher's materials, plans, worksheets and 30 photos, some side view, some vertical aerial and some oblique aerial. It covers a range of themes – land use, settlement, environmental and is set in the wider context of France.

Cost: £25.99 plus VAT (ISBN 0602 25955 X). Available from Ginn.

2 *European Locality Pack: Germany: Speyer on the Rhine, Key stage 2*
Another good pack on a similar basis at the French one above. Useful to fit with the rivers theme at some stage. Both these packs could be usefully supplemented by real maps of the area bought separately in good

bookshops, from Stanfords or obtained direct from the towns' tourist offices or the regional tourist offices.

Cost: £25.99 plus VAT (ISBN 0602 25956 8). Available from Ginn.

3 *Time and Place Geography: Resource Pack: Lansac (France), Key stage 2*
This is a very comprehensive pack based on a locality on the Aquitaine, a south western wine producing area of France. It includes high quality posters, 20 colour photographs, 2 Michelin maps, a vertical aerial view of the Dordogne/Garonne confluence, and a comprehensive teacher's guide with, of course, the inevitable photocopiable activity sheets. A real quality geography pack, but consequently expensive. If you are a Bordeaux wine enthusiast, it's fairly easy to supplement with wine labels, tourist information from the Bordelais, etc.

Cost: £60 approx. (ISBN 07501 0310 8). Available from Stanley Thornes.

4 *Montreuil: A European Place Study (France) (1995)*
Produced with key stage 1 and 2 in mind, this in-depth investigation of Montreuil explores the small market town in the Nord/Pas de Calais region. It contains 27 A4 colour photocards, a teacher's book, research and activity sheets.

Cost: £20 (members), £30 (non-members), ISBN 095 851 2881. Available from the Geographical Association (Co-ordinator Don Garman), see page 190 for address.

5 *Greece: Epidavros (1995)*
This small Greek fishing village can be studied through second-hand enquiry. The pack contains several large colour posters, 20 A4 colour photos, large colour maps and a range of teacher and pupil resources.

Cost: £49.99 (Folens ref. F631X). Available from Folens (address on page 174)

Books on whole countries (suitable for key stage 2)

As well as good locality pack resources, it is necessary to have good, up-to-date reference books which inform children about the wider area in which localities are situated. Some good ones are now being published.

Heinemann/Oxfam has a World Focus series which includes these titles to date:

- Bangladesh
- Brazil
- Kenya
- India
- Vietnam
- Ethiopia
- Jamaica
- South America.

It is a particularly good series and is cheap enough for a group set to be purchased. Each 32-page book also includes a 12-page case study of a locality and family.

Cost: £3.99 each.

Wayland have several series on countries:

1 *Countries of the World*
The publisher's notes claim for this series:

- A comprehensive introduction to different countries
- Highlights the distinctive features of each country enabling children to compare and contrast their lives with those of other lands and cultures.

Country	*ISBN*
Australia	0 7502 0260 2
Canada	0 7502 0261 0
The Caribbean	0 7202 0895 3
China	0 7502 0897 X
France	0 7502 0262 9
Greece	0 7502 1315 9
India	0 7502 0894 5
Italy	0 7502 1316 7

Japan	0 7502 0263 7
Pakistan	0 7502 1318 3
Spain	0 7502 1317 5
The USA	0 7502 0896 1

Cost: £4.99 each

2 *Our Country*
The publisher's notes claim for this series:

- A simple introduction to different countries through topics such as the weather, schools, homes and festivals.
- Includes short interviews with children from different backgrounds and parts of the country.

Country	*ISBN*
Australia	0 7502 0910 0
France	0 7502 0908 9
Greece	0 7502 1391 4
India	0 7502 1341 8
Japan	0 7502 0909 7
Pakistan	0 7502 1343 4
Spain	0 7502 1344 2
United Kingdom	0 7502 0911 9

Cost: £4.99 each

3 *The World's Rivers*
These recent river publications may also be useful for work on the rivers theme in the country of your chosen locality. The publishers claim for this series:

- Looks at some of the most important rivers in the world
- Looks in detail at the countries through which the river flows
- Discusses the environmental effects of the rivers.

River	*ISBN*
The Amazon	0 7502 1673 5
The Ganges	0 7502 1671 9
The Nile	0 7502 1674 3
The Thames	0 7502 1672 7

Cost: £3.99 each

The Geographical Association

The Geographical Association is a vital resource for primary school teachers via its publishing department. Primary and secondary classroom teachers, advisory teachers, inspectors and geographers involved in teacher training and in research institutes write voluntarily for the GA. Its publications seek to advise and inform on the teaching of geography in both primary and secondary education. It holds an annual three-day conference, of which at least one day is now targeted at primary teachers. You can become a personal member of the GA, or your school can become a corporate member. Publications are available more cheaply to members. Publications and membership details can be obtained from:

The Geographical Association
343 Fulwood Road
SHEFFIELD
S10 3BP
Tel: (0114) 2670666
Fax: (0114) 2670688

The following publications are of direct relevance to primary teachers. Coordinators should certainly consider purchasing copies of the first two for themselves and their school. New primary publications and resource packs are constantly being added.

Local Studies 5–13. Suggestions for the Non-specialist Teacher, updated 1991

Plans for Primary Geography, Wendy Morgan.

Geography, IT and the National Curriculum – Case Studies Booklet.

Geography Through Topics in Primary and Middle Schools Including the Application of Information Technology, 1989.

Primary Geography Matters: Inequalities Resources from Workshop and Keynote Lectures: Annual Conference, 1991.

Primary Geographer (a magazine produced four times a year, obtained automatically when you become a GA subscriber).

Teacher's books

A very useful and comprehensive compendium of primary geography resources is to be found in *Resources for key stages 1, 2 and 3*, Rachel Bowles, Geographical Association, 1993.

Primary Geography: A Developmental Approach, Jeremy Krause and Alan Blythe, Hodder & Stoughton, 1995.

The Outdoor Classroom: Educational Use, Landscape Design and Management of School Grounds (DES publication), Southgate Publishers.

Teaching Children Through the Environment, Hodder & Stoughton.

Using the School Surroundings, Stephen Scoffham, Ward Lock Educational.

Teaching Early Years Geography, Fran Martin, Chris Kington Publishing.

Teaching Geography at KS2, Bill Chambers and Karl Donert, Chris Kington Publishing, 1995.

Planning Primary Geography for key stages 1 and 2, Weldon and Richardson, John Murray, 1995.

World Studies 8–13, Simon Fisher and David Hicks, Oliver & Boyd, 1985.

Making Global Connections, David Hicks and Miriam Steiner (eds), Oliver & Boyd, 1989.

An Eye on the Environment, H.B. Joicey, Bell & Hyman.

Curriculum Leadership and Co-ordination in the Primary School: a Handbook for Teachers, Steve Harrison and Ken Theaker, Guild House Press, Guild House, Mitton Road, WHALLEY, Lancs.

Individually county or borough curriculum guidelines, geography sections.

Who's Who in the Environment: England (an excellent reference directory), The Environment Council, 80 York Way, LONDON, Tel: (0171) 8379688

Discovering Geology, Patrick H. Armstrong, Shire Publications.

Help for geography for special needs is given in *Geography for All*, Judy Sebba, David Fulton Publishers.

OFSTED/HMSO publications

Geography from 5–16, HMI Curriculum Matters Series (No. 7).

Environmental Education 5–16, HMI Curriculum Matters Series (No. 13).

Teaching and Learning of History and Geography, HMI Aspects of Primary Education Series.

These titles are available from:

HMSO Publications Centre
PO Box 276
LONDON
SW8 5DT
Tel: (0171) 8739090 (phone orders)
Fax orders: (0171) 8738200

HMSO publications are also available or can be ordered from provincial shops in certain towns and cities.

Magazines

Child Education and *Junior Education* (published monthly, sometimes contain articles relevant to geography).

Junior Projects (a bi-monthly magazine for Primary Teachers).

Primary file (a termly subscription service sometimes containing relevant articles).

Primary Geographer (see Geographical Association section).

Story books

The list indicates just a few of many books available to enhance geographical work.

Mainly for infants

The Lighthouse Keeper's Catastrophe, Ronda and David Armitage, Puffin Books.

Shaker Lane, Alice and Martin Provenson, Walker Books.

Little Red Riding Hood,: many versions are available.

Rainforest, Helen Cowcher, Andre Deutsch.

Antarctica, Helen Cowcher, Andre Deutsch.

Window, Jeannie Baker, Julia MacRae Books.

Dear Daddy, Philippe Dupasquier, Picture Puffins.

Morning Mollie, Shirley Hughes, Picture Lions.

Kim and the Watermelon, Miriam Smith, Picture Puffins.

The Journey Home, Joanne Findall, Walker Books.

The Great Round the World Balloon Race, Sue Scullard, Macmillan.

Little Polar Bear and the Brave Little Hare, Hans de Beer, North South Books.

Our House on the Hill, Philippe Dupasquier, Anderson Press.

Dinosaurs and All That Rubbish, Michael Foreman, Picture Puffins.

Beyond the UK for key stage 2

I am David, Anne Holm, Mammoth.

The Cay, Theodore Taylor, Puffin.

Hurricane, Andrew Salkey, Puffin.

Carefully chosen extracts from Laurie Lee's books for adults provide superb geographical descriptions of landscapes in the UK and Spain. Such extracts can be used successfully in the context of UK work and EU (Spain) work with older juniors. The books are not modern, but many of the descriptions still hold good, or provide a basis for discussing development and change in the landscape, rural and urban:

Cider with Rosie, Laurie Lee, Penguin

As I Walked Out One Midsummer Morning, Laurie Lee, Penguin

A Rose for Winter, Laurie Lee, Penguin.

Action plan for resources

1. Audit resources (see Resource Audit Sheet, see Appendix).
2. Prioritise collection and purchase needs.
3. Discuss budget for geography resources with head teacher within the School Development Plan context.
4. Obtain as many free posters, information packs, etc. as possible from the various organisations.
5. Involve pupils and staff in collecting postcards, pictures, photographs and other suitable materials.
6. Consider storage of resources. A central resource bank makes financial sense *but*

classroom-based resources make for most effective use with pupils.

Central resource bank

- Cheaper
- Is there a borrowing book?
- Who tidies it?
- A lot of time wasted fetching and returning items.

Classroom resources

- Instant access
- Permanent access
- Greater cost to school.

A compromise system is usually best, with masters OS maps and supplementary compasses centrally stored, for example, and atlases and some compasses in the classroom.

7 Always involve all or relevant staff when reviewing/assessing which published/commercial resources to buy.

8 Consider sharing more expensive resources if you work in a small group of schools. In theory, this is economic and helpful; but the practicalities of time and transport to fetch and return things need to be seriously considered. Fieldwork equipment, such as a soil augur or commercial clinometer, will not be used a lot. Control technology, such as a Roamer, could be a shared purchase by six rural schools used for a half-term in each school over the years. Similarly, very expensive equipment such as Weather Reporter, could be shared. It is better to plan use for half a term than not to have one at all.

9 Save yourself and your colleagues work by building up resource boxes or files on certain themes or localities. Ensure such resource boxes contain lists of their contents. Establish *who* is responsible for checking off the contents and returning them efficiently, otherwise hours of your time could be wasted. Consider photocopying any useful examples of pupils' work done relating to the unit or topic and including them in the box – they can be a help to staff who are new or uncertain as to what the outcomes the unit plans in schemes of work might look like. Key stage planning and keeping units of work within topics for several cycles of learning make for stability and less work in the end.

14

COORDINATING GEOGRAPHY IN THE SCHOOL

Whole-school plans

The context and timetable for the development of geography should be defined within the Whole School Development Plan, along with the plans for other subjects. There should be a section on the development of teacher assessment skills, techniques and moderation in geography within the whole school assessment policy (see Figure 14.1).

Figure 14.1

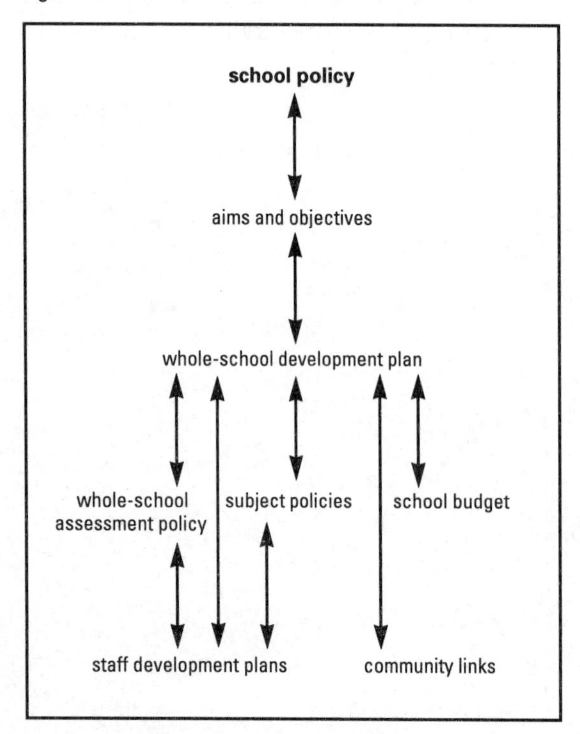

The role of the coordinator

As schools and school plans vary widely, so will the title and position of the person responsible for the teaching and learning of geography in the school. This person might be called coordinator, subject leader, subject manager, postholder, teacher responsible for geography or the unlucky teacher that was 'dumped' with geography! Throughout the book we have referred to the person in this crucial role as the coordinator because we feel it carries the important message that you work with other teachers, not on them, or at them, to bring about change.

The role of coordinator has sharpened in focus since the implementation of the geography Order. The role has developed from that of someone who knew where the topic resources were kept, what atlases and maps were in school, the geography of the local area, and so on, into a far more complex one:

- Developing the subject on all levels – skills, concepts, knowledge and understanding
- Designing a curriculum to fit the Revised National Curriculum requirements while maintaining a balance within the whole primary curriculum
- Guiding colleagues on what the docu-

ment really means, unit planning and teacher assessment.

Your main task as a coordinator is to see that, with, and through, your colleagues, the school develops good practice in primary geography teaching and learning.

All schools will be at different stages. You have to evaluate where you are, then move forward in small achievable steps, always keeping in mind an overview of your aim.

The role of coordinator can be broken down into five main areas:

- Communication
- Designing the curriculum
- Resources
- Assessment, evaluation and monitoring
- Staff development

Communication

Communication involves:

- Motivating, directing, aiding and assisting the whole school staff to develop geography teaching and learning
- Being a role model to encourage methods of good practice
- Sharing good work by spotting good displays in colleagues' rooms and redisplaying them in central places
- Labelling your own displays for children and adults
- Team teaching and group teaching with a colleague
- Helping other non-specialists
- Talking to parents and governors
- Cross-phase links

Designing the curriculum

Curriculum planning involves:

- Designing, directing and overseeing the development of a key stage plan
- Implementing and supporting the key stage plan
- Helping teachers to develop medium-term plans for units of work
- Advising on lesson plans and strategies
- Integrating IT into geography plans
- Integrating the cross-curricular themes into geography plans
- Developing a geography policy
- Developing a fieldwork health and safety policy
- Collecting SEN ideas for geography
- Key stage 1, 2 and 2, 3 liaison
- Monitoring teachers' plans.

Resources

Managing resources involves:

- Auditing the resources
- Consulting with colleagues on needs
- Organising the resources
- Checking new resources
- Budgeting for resources.

Assessment, evaluation and monitoring

These involve:

- Auditing where your school is now
- Advising on assessment
- Encouraging agreement trials
- Collecting a folder of moderated work as school examples
- Having an overview of the geography recording/records
- Possibly having a role in records for cross-phase liaison
- Monitoring children's achievement in geography in the school.

Staff development

The coordinator's role in staff development involves:

- Advising on staff INSET needs
- Attending courses to update and improve your knowledge
- Feeding information back from courses to the staff

- Working through school-focused staff development.

As the coordinator you will need to keep up to date with changes in National Curriculum documentation and geographical teaching to feed back to your colleagues. This will involve going on courses and inviting specialists into school to help you. Advisory services offer a range of courses on planning and teaching geography. There might also be a 5-, 10- or 20-day professional development course available through a university. It may be up to you to make sure that other teachers in the school are aware of any courses that would be of use to them, funds and the availability of local authority or independent training facilities being available.

The following are some areas you might like to consider for future INSET needs in your school:

- Planning from key questions
- Using open-ended IT programmes in geography
- Using specific geography IT programmes
- Using the concept keyboard for geography
- Key stage 2 geography teaching
- Progression in the use of the local area, through active learning, fieldwork, secondary sources and by drawing on pupils' own knowledge
- SEN and geography
- Progression in mapping
- Teaching distant places
- Differentiation in geography.

Sometimes if you are in a new school or new to the role you will need to have clarified for you:

- How much authority and responsibility the head teacher will offer you
- How much non-contact time, if any, the head teacher will be able to organise for you to develop your role
- How much INSET time and Grant for Educational Support and Training (GEST) funds will be available for geography
- Where geography fits within your appraisal targets and the whole-school development plan
- Whether the head teacher will enable you to *lead* through workshops, meetings, INSET activities, etc.
- What the (unlikely) scope is for possible specialist exchange and non-contact time to help you to develop the subject more thoroughly.

There is not room here to discuss in detail the skills and strategies needed by coordinators. Various books address this issue for those new to the role, or new to it in the context of the National Curriculum, and who would like to read more (see Chapter 13 and the Bibliography). Not all strategies and skills are unique to the role of geography coordinator and many can be applied in other primary curriculum areas, too.

Tackling the role

To help get yourself started, negotiate an action plan with the head based on the school's three-year development plan which should already have been developed as part of the wider school policy. Decide:

- What you have to do
- How you are going to do it
- Who might help you
- How long the job will take you (be realistic)
- How to prioritise the tasks (remember you have a class as well)
- On an action plan and write it.

Figure 14.2 suggests examples of what your action plan might look like.

Figure 14.2

Geography coordinator's action plan				April 97 Improving skills
Action	**How**	**By whom**	**By when**	**Further action**
Lead staff meeting on pre-mapping skills / progression in mapwork.	Collect mapwork examples from my class and any others, prepare some notes using Child Education and Primary Geographer articles Run workshop.	Co	July	Ask teachers to plan mapwork into next year's topics.
Ensure that Yr 2 children do mapping activities.	Check that Yr 2 teachers have mapping activities incorporated in "Our School" topic	Co with colleagues	Sept.	Ask for maps to be displayed in hall after half term.
Audit infant mapwork resources.	Teachers to use my audit sheet provided.	Colleagues	Mid Oct.	
Prioritise acquisition of new equipment.	Examine audit sheets.	Co	End Oct.	Pass list to headteacher - check it goes into school development plan eventually!
Review Yr 2 colleagues' mapwork.	Staff meeting sharing	Co and staff	End Nov.	Keep good examples of mapwork for school geography portfolio.

Geography coordinator's action plan				June 95
Action	**How**	**By whom**	**By when**	**Further action**
Audit geography in schools existing planning	Fill in a geography audit sheet for topics covered over last academic year.	Co	June	
Develop a Key Stage plan to implement Revised National Curriculum geography Order.	Draft plan Present to staff Support staff in its implementation	Co. + Dep. Head Whole Staff Co.	June July 12th Sept. →	 Amend plan if necessary Evaluate plan 1st year, amend if necessary, evaluate again.
Help staff to develop a key question approach to topic / unit planning	Go on planning course. School development day on developing medium term planning of geographical units	Co. AT + Co. Whole staff	Nov. Jan. 6th	Try to develop own unit Share with other staff Build up a bank of plans and work sheets to support Key Stage Plan.

Co. = Co-ordinator
AT = Advisory Teacher

Curriculum documentation

Why do we need a geography policy?

The reasons we need a geography policy are:

- To state the school's aims for implementing and developing geographical education for its pupils and to ensure it is carried out
- To inform teachers, new teachers, long-term supply teachers, governors, LEAs, OFSTED and parents of the ways in which this will be achieved
- To define long-term, medium-term and short-term planning objectives for geography
- To ensure that all members of staff are fully informed at any time of the school's geography policy
- So people know where they stand if verbal interpretations are unclear
- To provide a basis for review and evaluation at all scales of planning
- To ensure continuity of geography's curricular development within the school as staff changes occur.

What should go into a geography policy?

A geography policy should form part of the whole-school planning process and documentation. Once the overall aims, ethos and curriculum policy of the school have been clarified it is up to the geography coordinator, or a small planning team or the whole staff in a small rural school, to develop a policy for consideration by the whole staff. Their draft policy can be modified in the light of discussion with the whole staff. A sense of ownership of the school policies by the whole staff is important. They need to know why certain things have to be done in certain ways and they can often clarify issues for the policy developer who may be too close to the document to see an error or ambiguity.

The policy should not run to too many pages but be a short, concise statement of intentions, plans and ways to implement them which is usable by teachers – in short, a working document. Writing the policy will help the coordinator to think through and develop their own action plans to develop the subject over the next two to three years.

It should contain:

- The aims of teaching and learning in geography
- An action plan for the development of geography in the school
- The skills, concepts, attitudes and knowledge to be covered
- The teaching approaches to use in geography
  - the enquiry approach
  - key questions
  - fieldwork
  - issues
  - role-play and simulations
- A statement regarding the time allocation
- Integration of the subject into the whole primary curriculum
- Links with other subjects in the primary curriculum
- Long-term planning which shows how geography will be covered over a key stage
- Medium-term planning, units of work, a blank planning form and completed examples showing differentiation
- School assessment policy and marking schemes
- Recording techniques and the records required for pupils and teachers
- The place of cross-curricular themes and dimensions
- Fieldwork organisation, permission forms, health and safety requirements

- Reference to the school policies on equal opportunities, multi-cultural education and how they affect the teaching of geography
- Information about resources: amounts and location
- Special needs in geography
- Provision for evaluation of the planning and review of the document.

Evaluation of the policy

With so many changes happening in the primary curriculum, we have to be prepared to take a step backwards and review what is happening at all levels and stages in the teaching and learning of geography. Through the introduction of the National Curriculum substantial changes have needed to be made in the long-, medium- and short-term planning. These innovations need to be worked through, then reviewed and evaluated and adapted where necessary. As the whole staff will be involved in the implementation of the plans, they must also be involved in the evaluation. You can do this by reviewing different sections of the planning at different times. A teacher can quickly review a unit of work at the end and jot down any relevant comments such as 'Groups were too large', 'This key question worked well', 'This was a very heavy geography unit against a very heavy history unit'. Then, at the end of the year, the staff could get together and discuss how the units fitted into the whole primary curriculum and any changes that may need to be made.

The coordinator can collect and keep as a central resource the unit of work plans and any relevant work sheets. These resources can then provide a model for a teacher who has changed years, or a teacher new to the school. They do not need to be adhered to slavishly but will be useful as indicators of where things can be improved or changed. It means there is a starting point for medium- and short-term planning and we don't have to spend hours re-inventing the wheel. Traditionally teachers have enjoyed teaching new topics and units of work but we know that the precise planning required by the National Curriculum and teachers' work loads do not allow for this. Furthermore, it is a good idea to let planned units of work run for, say, three years or several cycles.

Year 1

- It's new – you run it through
- See how it could be improved
- Start collecting resources

Year 2

- Refine the units
- Feel more comfortable with them
- Sort out any snags within the overall plan
- The resource collection grows

Year 3

- Everything is now to hand
- There is little time spent on planning
- Look forward to minor changes or a different focus in the next year.

Any changes will obviously have to reflect national changes as well as school ones. With a structured plan it is possible to target the purchasing and collecting of new resources, which help in these times of financial constraints.

The success of a policy is visible, tangible evidence of it in practice in everyday school life, being internalised or as a working document – not a folder gathering dust on a shelf, to which only lip service is paid. The long-term final proof of the success of the policy will be in the pupils' enhanced achievements.

15

A ROUTE THROUGH OFSTED

The purpose of inspection

"The purpose of inspection is to identify strengths and weaknesses so that schools may improve the quality of education that they provide and raise the educational standards achieved by their pupils."

OFSTED, *Guidance on the Inspection of Nursery and Primary Schools* (HMSO, 1995)

The above quote is the premise on which the inspection system has been based. The revision to the framework for introduction in April 1996 has brought to the fore pupils' achievement and standards of education; with an emphasis on the progress made by pupils in a lesson, over a term, a year and their time at school. It takes into account the varying rates of progress made by pupils of different abilities, but for all pupils it presupposes a known starting point.

The composition of the final report to the school will change. How it is structured will depend on the size of the school, the way it organises its teaching and the extent of the evidence available. There may not necessarily be a specific section reporting individually on geography or other foundation subjects. Comments and findings about foundation subjects will be incorporated in the general sections of the report.

Providing evidence

Some weeks prior to the inspection the school will be asked to provide certain documents which will be used by the inspection team in the first stage of gathering evidence. These documents might include:

- School development plan
- Subject policies
- Subject guidelines where not included in the policy
- School policies on general matters, such as health and safety, assessment, possibly incorporated in the staff handbook
- Long-term key stage planning
- Medium-term planning, schemes of work
- Headteachers' form, a standard form produced by OFSTED which gives the team basic information on pupil numbers, finance, staffing and the general background to the school.

All inspectors use the documentation to gain an initial impression of the school, but also to raise issues particular to that school which they can investigate during the course of the inspection. Most inspectors have a fairly standard set of questions which they use to ensure they have covered all the necessary aspects in collecting their pre-inspection evidence. For a typical list of questions see Figure 15.1.

Figure 15.1

Pre-inspection analysis
• Is there a policy statement? • Does it include aims for geography? • Does it link geography to the whole curriculum? • Is it in line with whole school policies? • Does geography feature in the development plan? • Is there an identified co-ordinator? • Is adequate time allocated? • Is there a recurring experience for the subject? • Does the content meet NC requirements? • Do all pupils receive entitlement? • Is there integration of places, themes and skills? • Are there provisions for progression? • Is the sequence logical? • Is the place cover satisfactory ? • Are places studied at variety of scales? • Do pupils work outside the classroom? • Is there a safety statement? • Is an enquiry approach used? • Is there an indication of contribution to the key skills? • Do pupils use IT in geography? • Are there adequate resources? • Is there a geography portfolio?

Hopefully asking these questions would have produced evidence of the following areas or points to be clarified during the inspection from the headteacher, coordinator, teachers or pupils:

- Evidence of coordination and planning
- Evidence of NC coverage and match to the PoS
- Evidence of balance and progression
- Evidence of continuity between key stages
- Evidence of the enquiry approach
- Evidence of resources, finance
- Evidence of fieldwork opportunities
- Evidence of assessment
- Evidence of current targets or developments for geography
- Evidence of standards.

Shortly before the inspection teachers will be asked to provide a timetable showing what is happening in their classes during the days of the inspection. This is not an attempt to impose a secondary-type timetable but helps the inspectors to plan their visits to classes. Although one inspector on the team will be looking at geography, other inspectors may visit and look at geography particularly if two or three subjects are running at the same time on a carousel or if all the geography for that week is timetabled at the same time.

During the inspection

Early in a typical inspection, the inspector will have a conversation with the coordinator to clarify points outstanding from the pre-inspection evidence and ask general questions about management and finance which will contribute to the findings of the whole team on these aspects. The purpose is to allow the coordinator to help the inspector gain a clearer picture of the state of development of geography, its planning and its management within the context of the school. It may also allow an opportunity for the coordinator to indicate where they would find it useful to have comments from the inspector.

When inspectors visit classes for either a whole or part of a lesson, they will write comments and finally grade the lesson on an observation form. This form asks the inspector to consider four main areas:

- Teaching
- Response
- Attainment
- Progress.

In the context of the Observation form these are described by OFSTED below:

- *Teaching* Teaching should be considered primarily in relation to its impact on pupils' progress and attainment and the judgment should reflect this. Evidence should include those features which make teaching effective or not.

- *Response* Evidence here should focus on pupils' attitudes to the work in hand, their behaviour and any other aspects of their personal development which the lesson gives evidence of.
- *Attainment* Attainment is judged according to how well pupils are achieving in relation to national standards or expectations for their age. Evidence should be given of what pupils know, understand and can do. Text should highlight any significant variations among different groups of pupils.
- *Progress* Progress is judged according to how effectively pupils are acquiring and consolidating their knowledge, understanding and skills. Evidence should be given for this and for any significant variations among high, average and low attainers. In essence the question is 'what have the pupils gained from the lesson?'

When the inspector comes to your geography lesson, it is useful to have your normal lesson plan available, detailing if the pupils are working in groups, the rationale for those groups. Stick to your planned scheme of work, if this includes fieldwork so much the better. Depending on what is happening in the classroom, inspectors may sit and listen to the lesson, join a group, talk to children about their work or assess their general geographical knowledge. Don't expect a full feedback on each lesson, though the inspector may ask some questions and have a brief discussion, unless this disrupts the lesson. If you or any colleague want to talk to the inspector about geography tell them and arrange a mutually convienient time.

Features of a good geography lesson

Inspectors do not have a fixed blueprint for the perfect lesson. However, the best lessons seen have many of these features:

- Linked to medium- and long-term planning
- Clear learning outcomes
- Plans for this lesson clearly show progress from previous lesson and indicate likely direction of subsequent lessons
- Lesson has a distinct start and pupils are quickly aware of the purpose of the lesson, what is expected of them and what they are likely to cover
- Evidence of teacher's knowledge of the subject
- If teacher with whole class or small group, evidence of good use of questions to ascertain level of knowledge, develop enquiry, pose questions for enquiry
- Lesson has elements within it giving different experiences or working in different ways
- Lesson moves from one stage to another with suitable pace, but teacher needs to be aware of individual pupils' rates of work (is this reflected in the planning?)
- Pupils are confident in tackling tasks set, see the purpose of it and hopefully enjoy it
- Pupils cooperate with teacher and collaborate with other pupils
- Resources sufficient and suitable, at an appropriate level for the abilities of the pupils
- Pupils know how to use resources, including maps, atlases, globes, compasses, etc. (may be not all in one lesson)
- Lesson completed in time for a review and in time to clear away
- By the end of the lesson pupils can talk about what they have learned, possibly by reporting back to the group, teacher or whole class, depending on the organisation of the lesson
- Geography is taught through the medium of real places and defined localities.

Remember that an inspector can only report on what he or she sees. From your key stage plans and schemes of work you will know which classes will be doing geography during the inspection. You can find out from the teachers which aspects of the subject they are likely to be covering. It might be useful if you tell the inspector where he or she could see examples of previous work – does the school keep a portfolio of pupils' work? Most inspections will include a work sample where all the books showing the work of a small number of pupils will be inspected one evening after school. Depending on the term in which the inspection occurs these may show a lot or a little. If you are to be inspected in the autumn term, consider keeping examples from the previous year, but make sure it is clearly identified as such. This can be an excellent form of evidence to show progress.

Inspectors will also gather evidence from other sources (see Figure 15.2).

Figure 15.2

Inspectors will look at	To find
Geography portfolio	Standards of geography Evidence of progression Range of work and strategies Quality of marking
Wall displays	Is it enquiry based? Is it up-to-date? Is it children's work?
Geography textbooks and equipment	Is there enough ? Is it easily accessible? Are textbooks up-to-date?
Geography in the school library	Is there a geography section? Collection of different atlases ? Is it up-to-date ?
Staff meeting minutes	Has geography been discussed? Have decisions been implemented?
Effects of INSET	What has been covered? What evidence shows it has affected teaching and learning?
Pupils' records and reports	Is pupils' progress in geography recorded? Is the report just about attitude and coverage?
Individual education plans	Any reference to graphicacy? Are plans taken in to account when differentiating work?
Teachers' plans and notes	Do they identify clear learning objectives? Match pupil ability to tasks?

Feedback

The coordinator will have a scheduled oral feedback from the inspector. The inspector will use this to convey the initial impressions and to indicate clearly what the probable judgments will be with regard to geography. The comments will be about positive aspects, where things are going well, and indicate those areas where further work is necessary. It is important to take notes at this meeting of both the positive and negative features. Simply writing down what has to be done can make depressing reading and skews any feedback to the headteacher. If the coordinator feels that a criticism is being made because of oversight of evidence, now is the time to say. This discussion is likely to be your only opportunity for a professional dialogue with the inspector so make sure you make best use of the time available. Both the coordinator and the inspector will want to see geography in the school described accurately. The feedback time is also an opportunity to discuss possible strategies for improvement based on what the inspector has seen.

After the inspection

It is important that any external view of the school is seen in the context of improving the experiences of the pupils. Inspection is not an easy process but, if the school is prepared, is confident in knowing its own standards, then an external eye is useful in moderating those views and helpful in assisting the school to progress. If geography needs considerable attention, then it will need to feature in the school action plan. If you are doing well, then don't rest on your laurels; the encouragement should spur staff on to achieve even more in geography.

POSTSCRIPT

Where are we now?

What is the benefit of National Curriculum geography?

If National Curriculum geography disappeared tomorrow, what would we have learned?

How would we have benefited from its implementation?

It would have highlighted:

- The importance of the rigour of focusing on geographical concepts, skills and vocabulary
- The rigour needed when planning one subject within the whole primary framework
- That geography is not, and should not, be an isolated subject – during planning you are forced to isolate it and identify its links with the whole primary curriculum
- That careful planning, although time-consuming, actually *saves* time in the end because learning outcomes have been anticipated
- That an enquiry approach to learning is essential to geography
- That primary children can enjoy learning about people and places near and far
- That children of all ages are fascinated by maps and globes
- That good practice in primary geography, which existed in part in many schools before the National Curriculum, is spreading
- That fieldwork in geography is essential.

Good luck!

APPENDICES

Appendices A1, A2, A3 and A4 on pages 205–9 may be photocopied for use in your school. You will find that if you copy at 130% magnification you will have an A4 form.

Appendix A1

Key stage 1: Programme of study elements		
Programme of study	**Y1**	**Y2**
1a Investigate physical and human features		
1b Geographical questions: where? what? how?		
1c The wider world		
2 Enquiry: observe, question, record, communicate		
3a Use geographical terms		
3b Local fieldwork		
3c Follow directions		
3d Make maps, plans with signs and symbols		
3e Use globes		
Use maps and plans		
Identify key geographical features		
Locate/name countries of UK		
Locate home area on map		
Follow a route		
3f Use secondary sources		
4 School locality: school building		
School locality: school grounds		
School locality: local area		
Contrasting locality (UK or overseas)		
5a Main physical and human features of localities		
5b Similarities and differences between localities		
5c Effects of weather on people and places		
5d Land and building use		
6a Likes and dislikes of quality of environment		
6b Change in the environment		
6c Sustaining and improving environmental quality		

Appendix A2

Key stage 2: Programme of study elements					
Programme of study		**Y3**	**Y4**	**Y5**	**Y6**
1a	Investigate places and themes				
1b	Geographical questions: what? where? how and why?				
1c	Development of ability to recognise pattern, application of knowledge and understanding				
1d	Wider geographical context: links, range of scales				
2a	Enquiry: observe and question				
2b	Collect and record evidence				
2c	Analysis, conclusions and communication				
3a	Use geographical vocabulary				
3b	Fieldwork: instruments, techniques				
3c	Make maps and plans with symbols and key Use symbols and key				
3d	Use globes, maps, plans Co-ordinates Four-figure grid references Measure distance and direction Follow routes Use contents and index page of atlas Identify specified details: UK map (A) Identify specified details: Europe map (B) Identify specified details: World map (C)				
3e	Use secondary sources pictures, photographs, TV, visitors, books				
3f	Use IT for research and enquiry				
4	School locality Contrasting UK locality Contrasting locality in Africa, Asia (not Japan), South or Central America				
5a	Main physical and human features				
5b	Similarities and differences between localities				
5c	Effect of features of locality on human activity				
5d	Change in the locality				
5e	The broader context and links with other places				
6	UK context EU context World context Local scale Regional scale National scale				
7a	River system: features and catchment area				
7b	River processes: erosion and deposition				
8a	Weather: microclimate				
8b	Seasonal weather patterns				
8c	Weather round the world				
9a	Settlement: characteristics, location, economic activities				
9b	Land use in settlements, jobs				
9c	Land use issues				
10a	People's effects on the environment				
10b	Managing and sustaining the environment				

Appendix A3

Aide-memoire: Resource provision for primary geography (key stage 1)				
A well-equipped school will be likely to have for key stage 1:	**Central store**	**Library**	**Y1**	**Y2**
Globes: inflatable or other (1 per class)				
Playground compass rose				
A few direction compasses				
Infant big book atlases				
Infant individual atlases for Y2, e.g. Oxford Infant Atlas (Space to Great Britain), Collins Infant Atlas (Locality to the World)				
Play mats with geography features				
Standard 3D-maths shapes, Brio tracks, model houses and buildings for play activity and the development of plan view				
Range of wall/floor maps of UK, Europe and world, preferably made of easy clean materials				
Jigsaw maps of UK, Europe and world				
Range of published maps/plans, e.g tourist, high street				
Ordnance Survey 1:1250 scale map based on school				
And/or 1:2500 scale map based on school				
Plan of school site, permanent and consumable				
Pictorial maps: fictional and non-fiction				
Oblique aerial photos (can include postcards and photographs) of local area and distant places				
Large-scale vertical aerial photos of local area				
Side elevation photos (snapshots) of school site and local area including human and physical features				
Artefacts: rocks, fossils, clothes, products, sand, everyday objects from distant places				
Sufficient materials to support contrasting locality work, e.g videos, slides, commercial photopacks, charity photopacks with teacher's notes, pack of twinned school generated materials				
Simple reference books relating to countries, smaller localities and lifestyles				
Story books with a place content				
Large, easily-handled thermometers, rain gauges and wind direction equipment				

Appendix A4

Aide-memoire: Resource provision for primary geography (key stage 2)						
Many of the key stage 1 resources will also continue to be appropriate according to special educational needs.						
A well-equipped school will continue to regard as essential for key stage 2 from the key 1 resources:	**Central store**	**Library**	**Y3**	**Y4**	**Y5**	**Y6**
Ordnance Survey 1:1250 scale map based on school						
And/or 1:2500 scale map based on school						
Plans of school site, permanent and consumable						
Oblique aerial photos of local area and distant places						
Vertical aerial photos of local area and other localities						
Fiction or true stories with a real place context						
A well-equipped school will be likely to have for key stage 2 in addition to the above resources:						
Direction compasses (1 per class) and a half class set for fieldwork use is a useful minimum						
Atlases: a range in the school library a range in the class library at least a half set per class ideally, reasonably up-to-date						
Ordnance survey maps: 1:10 000 scale 1:25 000 scale (Pathfinder series) 1:50 000 scale (Landranger series) based on local area; and other localities to be studied						
Range of street maps, tourist maps, foreign maps, wall maps - UK, Europe, world and/or various countries/continents						
Reference materials relating to contrasting UK and economically developing locality or localities which exceed a few pages in a published text. Such resources may include:						
Videos						
Slides						
Maps/plans						
Snapshot photos/aerial photos						
Reference data and information provided by books, charity and commercial packs						

continued

Appendix A4 continued

	Central store	Library	Y3	Y4	Y5	Y6
Artefacts						
Pupils' work from a linked school						
Textbooks sets e.g Ginn Geography either as teacher resources or pupil resources						
Photocopiable worksheet resources for geography, commercial and teacher-produced						
The school library should have a reasonable quantity of up-to-date books relating to weather and climate, countries, water, environmental issues, industry, farming, transport, landscapes, rivers, towns and cities						
Tape measures: metre, surveyor's						
Trundle wheels						
Rain gauges						
Clinometers						
Stop watch						
Thermometers: range to include room, water/soil, maximum/minimum						
Wind speed gauge (anemometer)						
Wind sock or direction measurer						
Weather station						
Digital weather data logging equipment						

BIBLIOGRAPHY

Published by HMSO:

DES, *Geography in the National Curriculum* (HMSO, 1991)

DES, *National Curriculum Geography for Ages 5-16 Final Report* (HMSO, 1990)

DFE, *Geography in the National Curriculum* (HMSO, 1995)

DFE, *Art in the National Curriculum*(HMSO, 1995)

DFE, *English in the National Curriculum* (HMSO, 1995)

DFE, *History in the National Curriculum* (HMSO, 1995)

DFE, *IT in the National Curriculum* (HMSO, 1995)

DFE, *Mathematics in the National Curriculum* (HMSO, 1995)

DFE, *Music in the National Curriculum* (HMSO, 1995)

DFE, *Physical Education in the National Curriculum* (HMSO, 1995)

DFE, *Science in the National Curriculum* (HMSO, 1995)

DFE, *Design Technology in the National Curriculum* (HMSO, 1995)

DFE, *DATA: Guidance Materials for Design Technology: Key Stages 1 and 2* (HMSO, 1995)

HMI, *Matters for Discussion 5: The Teaching of Ideas in Geography* (HMSO, 1978)

HMI, *Geography from 5–16* (HMSO, 1986)

HMI, *Aspects of Primary Education: The Teaching and Learning of History and Geography* (HMSO, 1989)

HMI, *Mathematics Key Stages 1 and 2: A Report by HMI on the First Year 1989–90* (DES, 1991)

OFSTED, *Geography: A Review of Inspection Findings 1993/94* (HMSO, 1995)

OFSTED, *Inspection Issues and the Early Years: A Consultative Paper* (HMSO, 1995)

OFSTED, *Framework for the Inspection of Schools* (HMSO, 1995)

OFSTED, *Guidance on the Inspection of Nursery and Primary Schools* (HMSO, 1995)

Published by the Geographical Association:

Butt, G., *et al.*, *Assessment Works: Approaches to Assessment in Geography at Key Stages 1, 2 and 3* (Geographical Association, 1995)

May, S. and Thomas, T., *Fieldwork in Action 3: Managing Out-of-classroom Activities* (Geographical Association)

Mills, D. (ed.), *Geographical Work in the Primary and Middle Schools* (Geographical Association, 1988)

Milner, A., *Geography Starts Here* (Geographical Association, 1994)

Published by NCC:

NCC, *Economic and Industrial Understanding NCC 4* (NCC, 1990)

NCC, *Health Education NCC 5* (NCC, 1990)

NCC, *Careers Education NCC 6* (NCC, 1990)

NCC, *Environmental Education NCC 7* (NCC, 1990)

NCC, *Citizenship NCC 8* (NCC, 1991)

NCC, *An Introduction to Teaching Geography at Key Stages 1 and 2* (NCC, 1993)

Published by SCAA:

SCAA, *Pre-School Education Consultation: Draft Proposals* (SCAA, 1995)

SCAA, *Geography: Exemplification of Standards, Key Stage 3: Draft Materials* (SCAA, 1995)

SCAA, *Planning the Curriculum at Key Stages 1 and 2* (SCAA, 1995)

Other publications:

Blyth, A., and Krause, J., *Primary Geography: A Developmental Approach* (Hodder & Stoughton, 1995)

Harrison, S. and Thacker, K., *Curriculum Leadership and Co-ordination in the Primary School* (Guild House Press, 1991)

National Council for Educational Technology, *Focus on IT* (NCET, 1991)

Playfoot, D., Skelton, M. and Southworth, G., *The Primary School Management Book* (MGP, 1989)

Sutton, R., *Assessment: A Framework for Teaching* (NFER/Nelson, 1991)

TGAT, *Task Group for Assessment and Testing Report* (TGAT, 1988)

GLOSSARY

alternative technology small scale, simple technology that does not harm the environment, e.g. wooden ploughs drawn by oxen rather than pulled by tractor

anemometer an instrument to measure wind speed

barometer an instrument to measure air pressure

barrier a physical or mental obstacle to communication, e.g. mountain, safety fence, 'no go area'

'capes and bays' geography knowing facts, figures and names, without understanding of pattern and processes in the landscape

cay an area formed by the deposition of sand, e.g. Florida Keys

clinometer an instrument for measuring angles to calculate the slope or height of objects, such as trees or buildings

contour lines lines on a map which join points of equal height

cycle a series of events which happen repeatedly in a certain order

data any type of collected information on a given theme or topic

deposition the laying down of eroded or weathered material, e.g. a beach, river delta or mud bank

development education promotes greater understanding of the issues surrounding the relationship between less and further developed countries

Eckert's projection an elliptical map projection where the poles are shown by lines half the length of the Equator to minimise distortion of land area in the temperate latitudes

EDC Economically developing country

EU European Union

EIU Economic and Industrial Understanding

empathy appreciating the reasons for the actions of others in certain situations

enquiry questions a series of questions which lead pupils to investigate an issue or idea using primary or secondary sources

erosion the wearing away of rock and soil by ice, sea, rivers and wind to change the shape of the land

fieldwork work outside the classroom

GNP Gross National Product: the amount a country 'earns' in a year

graphicacy representing information in a non-written form, e.g. maps, diagrams, charts, tables, graphs

grid north the north imposed on a map by cartographers

grid a system of lines that are used to clarify, explain ideas or aid location

hazard perception the ability to predict the potential dangers of a site

igneous rock that was once molten within or on the earth's crust, e.g. granite, lava

'in the field' anywhere outside the classroom–school grounds, local street, etc.

INSET In-Service Education and Training

land use the way land is used by people to live or work, e.g. recreation, farming, housing, transport

locality a small area with distinctive features

location where something is to be found

magnetic north this is the north to which the compass needle points; it moves a fraction of a degree every year

matrix in the context of curriculum planning, a diagrammatic lined format on which you develop ideas. A table with columns so that links can be made between different factors, e.g. an environmental quality matrix

Mercator's projection a map projection where the lines of latitude get further apart towards the poles, so the northern hemisphere countries appear larger than they are

metamorphic igneous or sedimentary rocks which have been changed by heat and pressure, in the earth's crust, e.g. slate, marble

micro-climate the variations in climatic details on different parts of a site, e.g. small differences in temperature, and speed and humidity

migration the movement of people from one place to another

networks links between places in the landscape, e.g. bus route, telephone system, post person's rounds

OS Ordnance Survey

Peters projection a map projection showing countries according to their true area with accurate directions

pie chart a circle divided proportionally to show the values of the segments

primary industry obtaining raw materials e.g. mining, quarrying, fishing, farming, forestry

primary sources first-hand experience, e.g. fieldwork

process a series of events which cause a change

quadrat an area of land enclosed within a square (usually a metre) for data collection, especially of types of vegetation

region an area bigger than a locality which is defined by certain criteria, e.g. climatic conditions, position, landscape features, political definition

relief the height and shape of the earth's surface

secondary industry converting the products of primary industry into things that are useful through manufacturing, e.g. cars and paint

secondary sources pupils working from material collected by other people and presented in many forms

sediment material resulting from the weathering and erosion of the landscape

sedimentary rocks formed from sediments or shells often laid down in warm shallow seas, e.g. sandstone and limestone

settlement form the shape and layout of a settlement

settlement function the main human activities that take place in a settlement, e.g. industry, mining, shops, other services, transport and tourism

soil auger special giant screw-like tool for taking soil samples

spatial patterns patterns made in the landscape by places, features and people, e.g. transport, settlement, shopping

stack a stump of eroded rock in the sea, e.g. the Needles, Isle of Wight

systems separate parts which connect to make a whole, e.g. transport, river system

tertiary industry a service activity helping other industries and people, e.g. travel agent, dry cleaners, distribution warehouse

true north is the same as polar north, the North Pole on the globe

unit of study sequence of work relating to a particular topic, theme or issue

wave cut platform the area of the beach worn away and cleared by erosion, covered at high tide and sloping towards the sea

weathering the breakdown of a rock by the elements of weather, e.g. rain, frost, sun. Unlike erosion there is little or no movement

wide-angle-lens view study or overview of a region or country in which a locality is found

Winkelsche projection a map projection with a rounded grid so that directions are distorted, as are the shapes of countries on the edge of the map

zoom lens view study of the details of a locality

INDEX

Page numbers in **bold** indicate a main reference to the subject.